Bayesian Demographic Estimation and Forecasting

Bayesian Demographic Estimation and Forecasting

By
John Bryant
Junni L. Zhang

CRC Press
Taylor & Francis Group
Boca Raton London New York

CRC Press is an imprint of the
Taylor & Francis Group, an **informa** business

A CHAPMAN & HALL BOOK

Contents

Preface

This book presents new methods for answering questions such as

- How long can a newborn baby expect to live, given the ethnicity and income of the baby's parents?

- What proportion of future increases in health spending will be due to population aging?

- How should alternative estimates of population size be reconciled when these estimates do not agree?

The methods combine ideas from demography with ideas from statistics—especially a subfield known as Bayesian statistics.

The methods presented in this book form a coherent whole. We start with a simple framework, in which we model a single demographic series such as births or deaths. We then progressively expand the framework, to the point where we can deal with an entire system of interacting demographic series, measured through multiple, unreliable data sources.

Statistical demography is an interdisciplinary effort, involving statisticians and demographers, but also economists, epidemiologists, sociologists, and many others. We have tried to make the book useful and accessible to a broad audience. We assume only minimal knowledge of statistics, and no previous knowledge of demography. We focus on the underlying ideas, rather than the mathematical details.

We have developed a set of R packages implementing the methods in the book. The website for the book, www.bdef-book.com, has the packages along with data and code for all the examples. Most of the models could, alternatively, be fitted using general-purpose Bayesian computing environment such as BUGS or Stan. When fitting the models in this book, our packages are probably easier to work with, since they were designed specifically for the types of data and models used in the book. BUGS and Stan, however, offer more possibilities for extending the models.

The book is not a manual for our packages, and does not itself include any R code. Our hope is that readers wishing to find out about the new methods will use the book to learn about the principles, and will use the materials on the website to learn about the packages.

We focus on own particular approach to statistical demography, and make no attempt to review the field as a whole. Focusing on just one approach leads to a shorter and more unified book. However, statistical demography is a diverse, fascinating, and rapidly expanding field, and we point readers towards other approaches by giving suggested readings at the end of each chapter.

Acknowledgment

Jenny Harlow wrote the C code in our packages, and advised on software design. Jakub Bijak, Patrick Graham, Kirsten Nissen, Charlotte Taglioni, and Feifei Wang have worked with us on various aspects of Bayesian demography. The anonymous reviewers of the book gave numerous valuable suggestions. John Kimmel of CRC Press has been an ever-helpful guide during the writing of the book, and Robin Lloyd-Starkes has supervised the production process.

Statistics New Zealand has supported John Bryant in developing the software, and provided time and flexibility for working on the book. (The views expressed in this book are, however, entirely our own, and should not be attributed to Statistics New Zealand or to any other organization.) The Royal Society of New Zealand and the Center for Statistical Science and Guanghua School of Management at Peking University have funded several visits to Peking University by John Bryant, and the Guanghua School of Management has also funded a visit to New Zealand by Junni Zhang. We began our collaboration on Bayesian demography at the Small Area Estimation Conference 2013, and have received advice and encouragement from the small area estimation community since that time.

We thank our families for the love and support they have shown us during the writing of the book, and we dedicate the book to them.

1

Introduction

Goal 17.18 ... to increase significantly the availability of high-quality, timely and reliable data disaggregated by income, gender, age, race, ethnicity, migratory status, disability, geographic location and other characteristics relevant in national contexts.

— United Nations General Assembly

In September 2015, the United Nations General Assembly adopted the 2030 Agenda for Sustainable Development. The 2030 Agenda refers repeatedly to the need to compare progress made by smaller populations within the national population. Goal 1.1, for instance, emphasizes the importance of comparing poverty rates by sex, age, employment status, and geographical location (urban/rural). Goal 2.3 states that the productivity of small-scale food producers should be compared across sex, ethnicity, and occupation. Goal 17.18 calls more generally for the production of disaggregated statistics about development.

The focus on detailed comparisons in the 2030 Agenda is typical of contemporary analysis and policy making. Demand for disaggregated statistics is growing, not just among international organizations, but also among national governments, researchers, businesses, local authorities, and members of the public. Proponents of evidence-based policies want to target policies to tightly-defined groups. Strategists want to divide markets into ever-finer segments. Individuals want statistics about their own local area, age group, or ethnicity.

The demand for disaggregation extends to demographic forecasts. Users of demographic forecasts are no longer satisfied with national-level numbers. They want forecasts to be disaggregated by all the same variables that are used for estimates.

Producing reliable disaggregated demographic estimates and forecasts is difficult. Recent developments in statistical methodology have, however, opened up new possibilities. This book presents a new approach that combines ideas from a branch of statistics known as Bayesian statistics with ideas from mathematical demography. In this chapter, we describe some distinctive features of this new approach. But first, we illustrate our approach with an example.

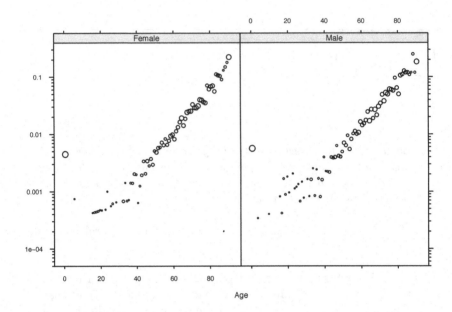

FIGURE 1.1: Estimates of Māori mortality rates in 2013 by single year of age. The graphs are drawn on a log scale. Each bubble shows the number of deaths divided by the corresponding population at risk. The size of each bubble is proportional to the number of deaths at that age. To preserve confidentiality, the deaths data have been randomly rounded to multiples of 3.

1.1 Example: Mortality Rates for Māori

Māori are the indigenous people of New Zealand. Measuring Māori progress towards lower mortality rates is a fundamental part of measuring New Zealand's overall social and economic performance.

Figure 1.1 shows estimates of the mortality rates for Māori in 2013, constructed by dividing the number of deaths in each age-sex group by the number of people in that group. In 2013, the Māori population of New Zealand was around 0.7 million, and there were just over 3,100 Māori deaths. The age groups in Figure 1.1 are single years, that is, age 0, age 1, age 2, and so on. The estimates are shown on a log scale. The diameters of the bubbles are proportional to the number of deaths: for instance, the bubble for 0-year-old males, for whom there were approximately 160 deaths is about 10 times larger than the bubble for 1-year-olds, for whom there were approximately 15 deaths.

New Zealand's official statistics agency, Stats NZ, has randomly rounded the deaths data to multiples of base 3, to protect confidentiality. Because the numbers of deaths in some cells are small, this rounding has a noticeable

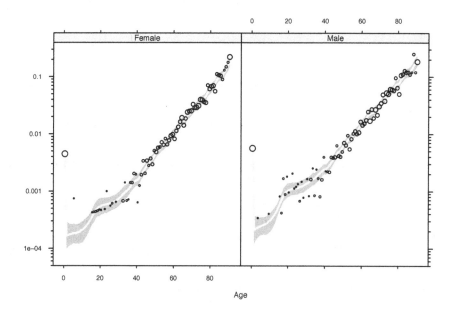

FIGURE 1.2: Model-based estimates of Māori mortality rates in 2012-2014 by single year of age. The graphs are drawn on a log scale. The gray bands are 95% credible intervals, and the pale lines are point estimates, from the model. The bubbles are simple estimates from Figure 1.1.

impact on the data. The apparent layering of mortality rates below age 40, particularly for males, is an artefact of the random rounding.

We would like to measure the underlying risk of dying faced by Māori in each age-sex group in 2013. The simple estimates shown in Figure 1.1 do not get these risks quite right. One reason is that random rounding distorts the rates. But the main reason is that observed number of deaths is affected by random variation, particularly when the numbers are small, which makes it a noisy indicator of underlying risks. In some years, for instance, some young age groups experience no deaths at all. We would not say that, in those years, the actual risk of dying was zero.

The models in this book provide ways of extracting signal from noisy data, so that we can see the underlying risks more clearly. Figure 1.2 shows results from one such model. We in fact fitted the model to data for 2006 as well as 2013 to boost sample sizes, but we show results only for 2013 here.

The gray bands in Figure 1.2 are 95% credible intervals. If the assumptions underlying the model are approximately correct, then there is a 95% chance that the true underlying rates lie within the credible intervals. Wider bands imply greater uncertainty. The pale lines in the middle of the credible intervals are point estimates, that is, single-number summaries of the results.

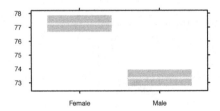

FIGURE 1.3: Māori life expectancy in 2013. The gray bands are 95% credible intervals, and the pale lines are point estimates.

The modeled estimates more or less follow a straight line from about age 30. Between the late teenage years and late 20s, the modeled rates are higher than we would expect from the straight line, particularly for males. This is the 'accident hump', a feature of most contemporary mortality profiles. The credible intervals are much wider for the young than for the old, reflecting the fact that there are fewer deaths, and therefore less data for estimating rates, for the young than the old.

Figure 1.3 shows estimates of life expectancy derived from the estimates of mortality rates. The uncertainty about the rates propagates through to the life expectancies. Here, as in the previous figures, the gray bands show 95% credible intervals.

The model used to generate the estimates in Figures 1.2 and 1.3 in fact does more than just estimate mortality rates. It also estimates the original, unrounded death counts, to remove any biases due to rounding. These estimates also come with measures of uncertainty. Table 1.1, for instance, shows estimates of the true number of deaths of 20 year old males. Once again, the answers are expressed in terms of probabilities. The table reports, for instance, that there is a 0.38 probability that the true, unconfidentialized number of deaths of 20 year olds was 6, and a 0.10 probability that it was 8.

TABLE 1.1
Estimated number of 20 year old males in 2013.

Value	4	5	6	7	8
Probability	0.08	0.22	0.38	0.22	0.10

1.2 Our Approach to Demographic Estimation and Forecasting

The methods in this book, like traditional demographic methods, focus on groups of people, as defined by characteristics such as age, sex, and region, rather than on individuals. Focusing on groups rather than individuals has disadvantages. For example, it limits the number of variables we can consider. But it also has advantages. It is well-suited to understanding change over time, and doing forecasting. It allows us to exploit datasets, such as confidentialized tables or historical time series, that have information on groups rather than individuals. And it matches the needs of many of the ultimate users of analyses. People planning schools and hospitals, or analyzing trends in mortality and population size, are interested in groups, not individuals.

We draw heavily on traditional demographic techniques for describing populations, from Lexis triangles (defined in Section 3.3), to origin-destination formats for migration (Section 4.3), to demographic accounts (Chapter 5). Compared with most traditional demography, however, the methods in this book are more formal and mathematical. We try to make our assumptions more explicit than demographers have traditionally done.

Taking a more formal approach requires making a clear distinction between (i) the true quantities that we wish to infer, and (ii) the data on which our inferences are based. For instance, we need to distinguish clearly between the true population counts that we are ultimately interested in, and the imperfect census data that are available to us for estimation.

Taking a formal approach also requires us to distinguish between (i) the underlying risks or propensities, and (ii) the random events governed by these risks or propensities. For instance, we need to distinguish between the probability that a woman aged 30 will give birth over the following year, and the actual proportion of women aged 30 who give birth.

Our approach to constructing formal statistical models draws on a branch of statistics known as Bayesian statistics. We say more about Bayesian statistics in Chapters 8 and 9, but it is an alternative to 'classical' statistics that has become increasingly influential across a range of disciplines. Bayesian statisticians are willing to use probabilities to characterize all forms of uncertainty, including uncertainty due to randomness, and uncertainty due to limited knowledge. Bayesian methods deal well with complex models, which makes them well suited to the problems of modern statistical demography.

The particular mix of mathematical demography and Bayesian statistics that we develop in this book emphasizes some features in particular:

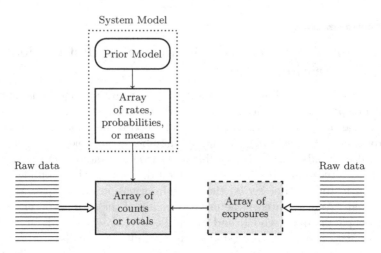

FIGURE 1.4: Inferring rates, probabilities, or means from a reliable dataset. The two gray rectangles are 'demographic arrays', constructed from raw individual-level data. The gray rectangles are partly or fully observed, and are treated as if they were completely error-free. We use the array of counts or totals to infer the array of rates, probabilities, or means, with the help of the prior model. The array of exposures is optional.

1. Disaggregation

2. Forecasting

3. Unreliable data

4. Demographic systems

As we will discuss in Section 5.1, a demographic system is a set of interrelated demographic series, such as population, births, deaths, and migration. Some demographic applications involve all four features—disaggregation, forecasting, unreliable data, *and* demographic systems—but others do not. We present three general frameworks, aimed at progressively more complicated applications. The first framework, which is the simplest, is summarized in Figure 1.4.

The framework is used when estimating underlying rates, probabilities, or means from data on counts or totals. The data and the underlying rates, probabilities, or means are organized in the form of demographic arrays. 'Demographic array' is our term for counts, rates, or other values cross-tabulated by dimensions such as age, sex, and time.

We assume that the data to hand are sufficiently reliable that they can be treated as if they were completely error-free. In other words, we pretend that the reported values for births, deaths, population, or any other demographic quantity are identical to the true values. This assumption sounds extreme when stated in this way, but it is in fact standard in much statistical modeling.

It is a sensible strategy when the errors are small, and have only a minor effect on results.

The framework of Figure 1.4 does, however, allow for the possibility of gaps in the data. Within the framework, the raw data, and hence the arrays of counts or totals that are assembled from this data, may not have values for every combination of age, sex, time, or other dimensions that we are interested in. When values are missing, we infer them as part of the overall estimation process.

Because the framework allows for missing data, it allows for demographic forecasts. In a forecast, the data has the format

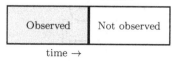

time →

We have data up to a certain time, and then no data past that time. Consider, for instance, the problem of forecasting birth rates for 2019–2028, based on data for 1990–2018. One way of formulating the problem is that we are estimating birth rates for the combined period 1990–2028, but have missing data for 2019–2028.

When estimating underlying rates, probabilities, or means it is usually wise to take into account regularities in the data, such as a shared tendency for death rates to rise with age. Similarly, we may be able to improve our estimates by incorporating information from other settings, such as information on plausible ranges. Strengthening estimates through the sharing of information is the job of the prior model. In the framework of Figure 1.4, the prior model and the model describing how counts or totals are generated given rates, probabilities, or means are referred to jointly as the 'system model'.

When modeling events, we generally need to take account of the size of the population that is at risk of experiencing the event. When modeling deaths by region, for instance, we would usually want to include a measure of regional population size. We refer to measures of population at risk as 'exposures'. In the framework of Figure 1.4, we treat exposures as error-free.

Although the framework of Figure 1.4 has wide application, sometimes the measurement errors in the data are *not* small, and ignoring them *would* materially affect the conclusions from the analysis. In such cases, we need to shift to the framework of Figure 1.5.

The key difference between the frameworks of Figures 1.4 and 1.5 is that in Figure 1.5 the true array of counts or totals is no longer treated as directly observed. Instead, all that is observed is one or more arrays of unreliable data. We must infer true counts or totals from the unreliable data.

To do so, we have to make some assumptions about the nature of the unreliability of the data. These assumptions are captured by 'data models'. Each dataset has one data model. The data model expresses how the dataset is generated given the true counts or totals. A simple data model, for instance,

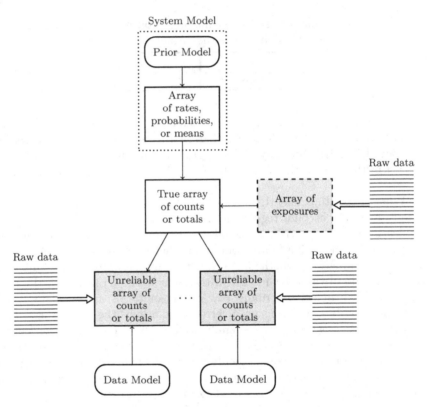

FIGURE 1.5: Inferring rates, probabilities, or means, plus the associated true counts or totals. In contrast to the framework of Figure 1.4, the array of counts or totals has to be inferred from one or more unreliable datasets, which requires specifying models for the relationship between the true counts or totals and the data.

might say that a dataset captures 50% of the true events, so that if 20 births occur during a period, approximately 10 of these will appear in the dataset.

The system model is completely unaffected by the change in the status of the array of counts or totals. In the framework of Figure 1.5, just like that of Figure 1.4, the system model is composed of the model describing how true counts or totals are generated given rates, probabilities, or means, plus a prior model capturing expected trends and patterns.

The exposure term in the framework of Figure 1.5 is also identical to that of Figure 1.4. It is optional, but if included, is treated as error-free.

The frameworks of Figures 1.4 and 1.5 both deal with single demographic series. With the framework of Figure 1.6, we move to looking at an entire demographic system.

Instead of a single array of counts or totals, we have multiple arrays of counts. Each array represents a demographic series. There is always an array containing population counts. The remaining arrays contain counts of events such as births, deaths, and migrations that specify how people enter, exit, or move around the system. These counts are treated as unobserved, and must be inferred as part of the estimation process.

The arrays of counts together form a demographic account. The elements within a demographic account conform to a fundamental accounting identity: population at the end of each period equals population at the beginning plus entries minus exits.

The system model for each series in the demographic account is essentially the same as the system models in the two previous frameworks. The main difference is that exposure is not treated as fixed and known. Instead, it is derived from the array of population counts during estimation.

Each array of counts can be described by one or more datasets, and each dataset has its own data model. An array can even have no datasets, in which case it is inferred from the other arrays.

1.3 Outline of the Rest of the Book

Part I covers the demographic aspects of our approach: demographic arrays, demographic accounts, and their relationship with individual-level biographies. Part II covers the Bayesian statistical methods that we use to infer the unknown quantities in Figures 1.4–1.6.

Part III looks at models falling within the framework of Figure 1.4, where we have a single array of counts or totals that we treat as error free. Part III starts with a short chapter where we describe the framework in more detail. There are then three applications chapters, each presenting a case study based on the framework. Parts IV and V deal with the frameworks of Figures 1.5 and 1.6. They follow the same introduction-applications format as Part III.

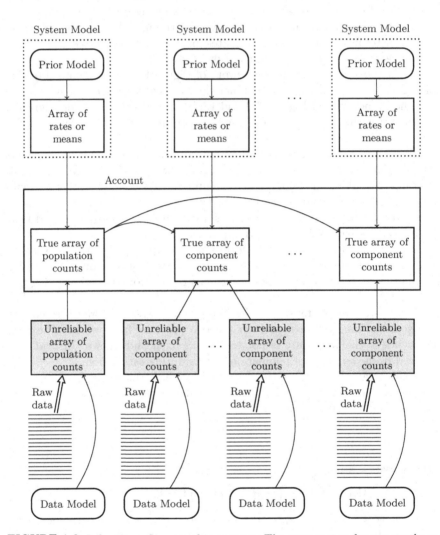

FIGURE 1.6: Inferring a demographic account. The gray rectangles are partly or fully observed; everything else is unknown and needs to be inferred. Each series in the demographic account can be measured by one or more datasets, or may not be measured by any dataset. Each dataset has its own data model. Each demographic series has its own system model.

We conclude the book with some reflections on the future development of the methods.

1.4 References and Further Reading

The opening quote and description of the goals of the 2030 Agenda for Sustainable Development come from United Nations General Assembly (2015).

The data on Māori mortality rates come from the table *Deaths by age and sex (Maori population) (Annual-Dec)* from the *Infoshare* database on the Stats NZ website. The data on Māori population come from the table *Estimated resident population (ERP), national population by ethnic group, age, and sex, 30 June 1996, 2001, 2006, and 2013* from the *NZ.Stat* database on the Stats NZ website. The data were downloaded on October 24, 2016. The code and data for the model are available at the website for this book, www.bdef-book.com.

Figures 1.1–1.3 were drawn using functions from *R* package **lattice** (R Core Team, 2016; Sarkar, 2008). Figures 1.4–1.6 were drawn using the **TikZ** package (Tantau, 2008). We use **lattice** and TikZ extensively through the rest of the book.

Introductions to demographic methods include Preston et al. (2001) and Wachter (2014). Demographic texts with a greater focus on estimation include Smith et al. (2013) and Swanson and Tayman (2012). Alho and Spencer (2005) is a wide-ranging textbook on statistical demography. Bijak and Bryant (2016) review Bayesian demography.

In 2015, the United Nations Population Division used Bayesian methods for its global population projections (Gerland et al., 2014; Raftery et al., 2014; UN Population Division, 2015). Abel et al. (2016) critique the projections.

Gelman and Hill (2007); Lynch (2007) are textbooks on Bayesian statistics aimed at social scientists, and Gelman et al. (2014) and McElreath (2016) are texts on Bayesian statistics in general.

Part I

Demographic Foundations

2

Demographic Foundations

The process of demographic measurement and modeling can be distilled into the four steps listed in Table 2.1. Data are collected on people and events. The data are put into standard formats. Demographic measures such as rates, probabilities, or means are estimated. The measures are summarized, using indicators such as growth rates, population totals, or life expectancies.

TABLE 2.1
The process of demographic measurement and modeling

Step	Description	Example
1	Collect data on individuals	Register deaths; count people
2	Organize data	Tabulations of deaths, exposure
3	Estimate	Calculate mortality rates
4	Summarize	Calculate life expectancy

Our main aim in this book is to introduce new methods for carrying out step 3, estimation. But to do step 3, we need to understand steps 1 and 2. In Part I, we review these earlier steps.

The two core concepts of Part I are demographic arrays and demographic accounts. A demographic array is a set of cross-tabulated counts, rates, or other demographic quantities. The term 'demographic array' is not standard in demography, but the concept is. Anyone who has ever opened a demographic textbook, or worked with demographic data, will have encountered these sorts of tabulations. They are the standard way of formatting the inputs and outputs of a demographic analysis.

Demographic accounts are the demographic equivalent of national accounts. A demographic account is a collection of linked demographic arrays. The elements of arrays within an account conform to a set of fundamental accounting identities: for every subpopulation, population count at the end of each period equals population count at the beginning of the period, plus count of entries during the period, and minus count of exits during the period. Demographic accounts were first developed several decades ago, but have not received the wide use that, in our opinion, they deserve.

Demographic arrays and accounts are constructed from individual-level biographies. The construction process is selective, retaining some aspects of the individual-level biographies, and omitting others. It also entails a shift in perspective, from individuals to groups.

We illustrate the process of constructing arrays and accounts using a population of 12 fictitious individuals. We introduce the 12 individuals in Chapter 3, construct arrays describing their lives in Chapter 4, and combine arrays to form accounts in Chapter 5. To construct the arrays, we need a number of demographic concepts such as Lexis triangles, exposures, and cohorts, which we discuss as they arise.

In reality, individual biographies are not collected perfectly. Correspondingly, arrays have measurement errors and gaps, and accounting identities are not satisfied. Chapter 6 briefly looks at demographic data and the modeling choices.

2.1   References and Further Reading

Willekens (2006) presents a framework for mathematical demography that connects individual-level events to population-level processes. Courgeau (1985) is a classic article on the connection between aggregate demographic events and individual life courses.

Rees and Wilson (1977) and Stone (1984) were early proponents of demographic accounts. The latter reference is the acceptance speech given by Stone when he received a Nobel Prize in economics, for his contribution to the development of national accounts. Stone argues in his speech that countries should develop social and demographic accounts to complement economic accounts.

In computer science terminology, demographic arrays and demographic accounts are types of data structure (Wegner and Reilly, 2003). In the R packages implementing our approach, demographic arrays and accounts are literally data structures, with explicit rules on the data they can contain and the ways they can be manipulated.

Preston et al. (2001) is an authoritative textbook on traditional demographic methods.

3

Demographic Individuals

Demography is the statistical study of populations. Data on populations are, however, assembled from data on individuals. Demographers have their own distinctive way of representing individuals, and converting their biographies into data.

3.1 Attributes

When describing individuals, demographers concentrate on a handful of attributes. These attributes typically include age and sex, but, depending on the application, can include others, such as region of birth, region of current residence, ethnicity, marital status, educational status, health status, or labor force status.

At the individual level, age can treated as a continuous variable. If an individual's date of birth is known, then the person's exact age can be calculated at any date required. Demographers treat all other attributes as consisting of a limited number of statuses. If the attribute is marital status, then the statuses might be "Unmarried", "Currently married", "Divorced/separated", and "Widowed". If the attribute in question is educational status, for instance, then the statuses might be "None", "Secondary or less", and "Tertiary".

Some attributes are necessarily fixed over a person's lifetime. A person's region of birth, for instance, never changes. Other attributes can change. A person's region of current residence typically changes over the person's lifetime, for example, as does marital status, educational status, and health status. A few attributes are treated as fixed in some applications and changeable in others. Some studies treat ethnicity as fixed at birth, for instance, while others treat it as something that people change as their identities evolve.

3.2   Events

Births, deaths, and marriages are all examples of individual-level events. In this book, we distinguish between three types of individual-level events:

1. Entries and exits

2. Movements between statuses

3. Non-demographic events

Entries bring individuals into the population of interest, and exits take them out. If the population of interest is the population of a country, for instance, then births and immigrations would count as entries, while deaths and emigrations would count as exits. Entries increase population size, and exits decrease population size.

The classic example of movements between statuses is internal migration between different regions of a country. However, changes to ethnicity, marital status, and labor force status are also types of movements. Movements between statuses do not affect overall population size. They do, however, affect population structure—that is, they affect the way the population is distributed across the various groups.

Aging is a type of movement between statuses. It is, however, a special type of movement, in that, provided the individual person remains alive, the movements are completely predictable.

Non-demographic events do not directly affect population size or structure. Examples include the receipt of income, expenditure on health care, and hours worked. The process of earning income, for instance, can be represented as monthly events in which a person is paid a salary. Health expenditure can be represented as an series of payments to healthcare providers. Employment can be represented as a series of work episodes.

With demographic events, we simply record that an event occurred. With non-demographic events, we typically record an amount or quantity. We record the amount paid to the healthcare provider, for instance, or the number of hours worked.

The idea of a non-demographic event is not standard in demography, but it is something that we have found useful in our own work. As we will see in Chapter 13, allowing for non-demographic events permits us to treat income, health expenditures, and hours worked within the same framework as core demographic topics such as births, deaths, and migration.

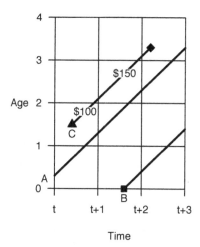

FIGURE 3.1: A Lexis diagram showing lifelines and events for individuals A, B, and C. The square denotes a birth, the triangle denotes an immigration, the diamond denotes a death, and the dollar amounts denote medical expenses.

3.3   Lexis Diagram

A Lexis diagram is a visual device for reasoning about the relationship between age and time, and the relationship between events and population size. Lexis diagrams can also be used to depict the features of people's lives that are captured by demographic models.

Figure 3.1 is an example of a Lexis diagram. The horizontal axis shows time and the vertical axis shows age. The diagonal lines marked A, B, and C are lifelines for individuals A, B, and C. Lifelines show individuals' exact ages at each point in time. The lifelines all slope upwards at 45 degrees, reflecting the fact that an individual gains precisely one year of age with each calendar year. The square, triangle, diamond, and dollar signs all represent events.

From the point of view of a demographer modeling fertility, mortality, migration, or medical expenditures, Individual A's life between times t and $t + 3$ is extremely simple. At time t, she is 0.3 years old. At time $t + 3$ she is 3.3 years old. Between t and $t + 3$ she experiences no events.

A demographer would record that Individual B was born at time $t + 1.6$, but from then on would record only that individual B survived through the rest of the observation period.

Individual C yields more data. She joins the population at time $t + 0.4$ through immigration, aged 1.5. At time $t + 0.7$, she has a medical expense of $100, and at time $t + 1.6$ has a medical expense of $150. At time $t + 2.2$, she dies aged 3.3.

3.4   Twelve Fictitious Individuals

We conclude this chapter by summarizing the lives of the 12 fictitious individuals who will supply the data for the next two chapters.

The individuals are all female, and live in a country composed of two regions called East and West. We observe the ages and regions of the 12 individuals in year 1980, and follow them over the period 1980–2000, keeping track of births, deaths, migrations, and payment of taxes. For each event, the year, the age of the person experiencing the event, and the region where the event occurred are recorded.

The life of our first individual, Anna, is depicted in Figure 3.2. In 1980, Anna is living in East and is 10 years old. She stays in East until 1998, when, aged 28, she migrates to the West. She returns to East in 2003 at the age of 33, and then, in 2006, at the age of 36, she has a baby. In 2010, Anna is 40 years old and still living in East.

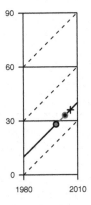

FIGURE 3.2: The life of Anna, summarized by a Lexis diagram. The **O** symbol represents a movement from East Region to West Region, and the **●** symbol represents a movement from West Region to East Region. The **+** symbol represents a birth.

Figure 3.3, which extends over 3 pages, uses Lexis diagrams to summarize the lives of all 12 individuals.

—	—	In East/West
■	▪	Born in East/West
◆	◆	Death in East/West
▲	▲	Immigration to East/West
▼	▼	Emigration from East/West
○	●	Internal migration from East to West / West to East
+	+	Giving birth in East/West
$50	$50	Tax payment in East/West

FIGURE 3.3a: Symbols used in Lexis diagrams for the 12 fictitious individuals. The Lexis diagrams are displayed on the following two pages. The table above gives the legend. Lifelines or events where the individual is living in East Region are shown in black, and lifelines or events where the individual is living in West region are shown in gray.

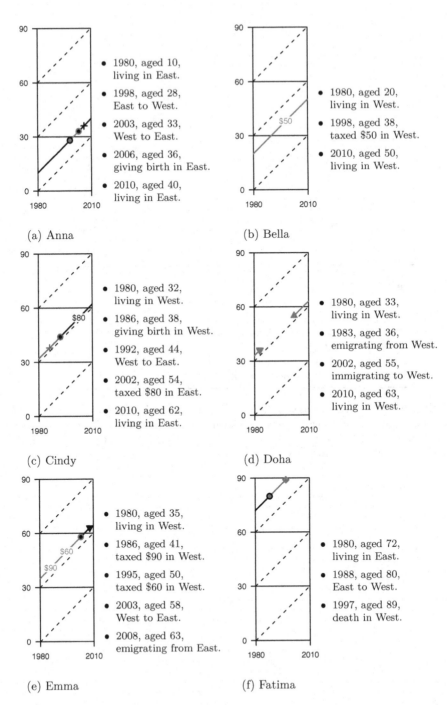

(a) Anna

(b) Bella

(c) Cindy

(d) Doha

(e) Emma

(f) Fatima

FIGURE 3.3b: Lexis diagrams and biographies for the first 6 individuals.

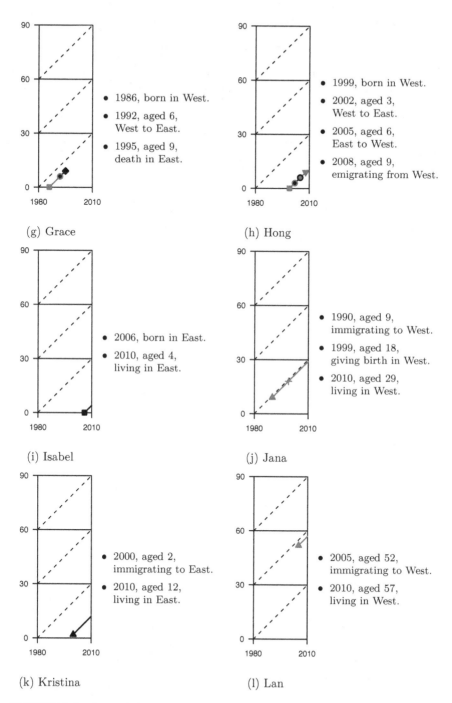

FIGURE 3.3c: Lexis diagrams and biographies for the remaining 6 individuals.

3.5 References and Further Reading

Preston et al. (2001, pp. 32-33) describe the Lexis diagram. Lexis diagrams are named after a 19th century demographer, but, consistent with the law that no scientific discovery is ever named after its actual discoverer, it turns that he was not the first person to use them (Stigler, 1980; Vandeschrick, 2001). Alho and Spencer (2005) review a variety of individual-level statistical models used by demographers.

4

Demographic Arrays

'Demographic array' is our term for cross-tabulated counts, rates, or other values used in group-level demographic models. A table of population counts arranged by age and sex, for instance, is a demographic array, as is a table of death rates by age, sex, ethnicity, and time.

The dimensions of a demographic array depend on the individual-level attributes that were measured, plus, in most applications, time. A demographic array contains one cell for every possible combination of the dimensions: one cell, for instance, for every possible combination of age, sex, and time. When the number of dimensions, or the number of categories within each dimension, is large, a demographic array can contain thousands, or hundreds of thousands, of cells.

In this chapter, we look at how individual-level data are turned into demographic arrays, and at how these arrays are combined or manipulated.

4.1 Population Counts

We start with population counts for the 12 individuals from Chapter 3. The process of constructing an array of population counts is illustrated in Figure 4.1.

Our basic inputs are the name, age, and location of everyone who was in the country in 1980 or 2010, as shown in Panel (a). The first step is to remove the names, and convert exact ages to age intervals, yielding the data in Panel (b). The removal of the names is significant. It illustrates the fact that demographic models work with groups and not individuals. All we pay attention to, in group-level demographic modeling, is the group that each person belongs to. Once we know that, the names are irrelevant.

Panel (b) of Figure 4.1 uses age intervals rather than exact ages. By replacing exact ages with age intervals we are throwing away information. For instance, we are obscuring the fact that Cindy, Doha, and Emma are much younger than Fatima. The 30-year intervals of Panel (b) are wider than the intervals used in most demographic analyses. But even 1-year or 5-year intervals are less informative than exact ages. Given the importance that demographers place on age, why use intervals?

Name	Year	Age	Region
Anna	1980	10	East
Bella	1980	20	West
Cindy	1980	32	West
Doha	1980	33	West
Emma	1980	35	West
Fatima	1980	72	East
Isabel	2010	4	East
Kristina	2010	12	East
Jana	2010	29	West
Anna	2010	40	East
Bella	2010	50	West
Lan	2010	57	West
Cindy	2010	62	East
Doha	2010	63	West

Year	Age	Region
1980	0–29	East
1980	0–29	West
1980	30+	West
1980	30+	West
1980	30+	West
1980	30+	East
2010	0–29	East
2010	0–29	East
2010	0–29	West
2010	30+	East
2010	30+	West
2010	30+	West
2010	30+	East
2010	30+	West

(a) Raw data   (b) Processed data

	1980		2010	
	West	East	West	East
0–29	1	1	1	2
30+	3	1	3	2

(c) Array

FIGURE 4.1: Constructing an array of population counts in 1980.

The reason for using intervals is that we are allocating people to groups. If we were to use all the age information available to us, and, for instance, distinguish people who were one day old from people who were two days old, three days old, and so on, then our datasets would have many more groups, each of which was tiny.

Getting from Panel (b) to Panel (c) in Figure 4.1 is easy. We count the number of people in each possible combination of time, age, and region, and record the number in the corresponding cell of the array.

4.2 Death Counts

Next we construct an array of death counts. The raw data, shown in Panel (a) of Figure 4.2, are very similar to the raw data for population counts. However, the processed data for deaths, shown in Panel (b), differ from the processed data for population counts in an important way: they use periods rather than exact times. This is an example of a more general principle, summarized in Figure 4.3, that population counts are measured at exact times, while events are measured over periods.

Name	Year	Age	Region
Grace	1995	9	East
Fatima	1997	89	West

(a) Raw data

Period	Age	Region
1980–2000	0–29	East
1980–2000	30+	West

(b) Processed data

	1980–2000	
	West	**East**
0–29	0	1
30+	1	0

(c) Array

FIGURE 4.2: Constructing an array of death counts.

Forgetting that counts of population and counts of events have different time references is a common source of confusion in applied demography. A particularly confusing feature of demographic data is that there is usually one more count for population than there is for events, since the exact times enclose the periods. In Figure 4.3, for instance, population is counted three times, while events are counted twice.

Going from Panel (b) to Panel (c) in Figure 4.2 is again a matter of adding up within each possible combination of time, age, and region. If a particular

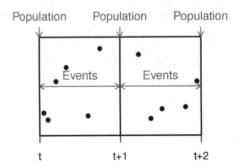

FIGURE 4.3: Population counts are measured at exact times, and events are measured over periods. Here, for instance, population is measured at exact times $t, t+1$, and $t+2$, while events are measured during the interval between t and $t+1$ and the interval between $t+1$ and $t+2$.

combination has no events, then the corresponding cell in the array has a value of 0. In Figure 4.2 there are no deaths of 0–29 year olds in West and no deaths of 30+ year olds in East. The associated cells in the demographic array therefore both have values of 0.

4.3 Movements

Movements, such as internal migration between regions, are special types of events, in that they involve an origin and a destination. Arrays of movements are, however, constructed in essentially the same way as other arrays. Figure 4.4 shows an example.

A distinctive feature of origin-destination arrays is that they can contain 'structural zeros'. The array in Panel (c) of Figure 4.4, for instance, contains structural zeros along the diagonal. A structural zero is a logical consequence of the way a problem is set up. The structural zeros in Panel (c) are a consequence of the fact that we are measuring movements between regions, but are not measuring movements within each region. When we do not measure movements within each region, the count of movements from a region to itself is necessarily zero.

Name	Year	Age	Origin	Destination
Fatima	1988	80	East	West
Grace	1992	6	West	East
Cindy	1992	44	West	East
Anna	1998	28	East	West
Hong	2002	3	West	East
Anna	2003	33	West	East
Emma	2003	58	West	East
Hong	2005	6	East	West

(a) Raw data

Period	Age	Origin	Destination
1980–2000	30+	East	West
1980–2000	0–29	West	East
1980–2000	30+	West	East
1980–2000	0–29	East	West
1980–2000	0–29	West	East
1980–2000	30+	West	East
1980–2000	30+	West	East
1980–2000	0–29	East	West

(b) Processed data

		1980–2000						
	0–29					**30+**		
		Destination					Destination	
		West	East				West	East
Origin	West	0	2		Origin	West	0	3
	East	2	0			East	1	0

(c) Array

FIGURE 4.4: Constructing an array of internal migration counts.

4.4 Alternative Representations of Changing Statuses

The origin-destination format is a natural way to describe movements between statuses. It is also an attractive format for model-building, since it allows models to exploit information on the origin, on the destination, and on any special relationships between the two. However, even with a modest number of statuses, an array in origin-destination format can be prohibitively large. With 10 statuses, and no other dimensions, an origin-destination array has 100 cells, but with 1,000 statuses, it has 1,000,000.

Demographers have, accordingly, developed more parsimonious formats for describing changing statuses. One example is the pool format, shown in Figure 4.5. With the pool format, total outward movements and total inward movements are shown for each status, but without any information linking origins to destinations.

By definition, the total number of outward movements has to equal the total number of inward movements, within each combination of the other dimensions. For instance, in Figure 4.5, the total number of outward movements for 0–29 year olds must equal the total number of inward movements for 0–29 year olds, and the total number of outward movements for 30+ year olds must equal the total number of inward movements for 30+ year olds.

The pool format retains some of the interpretability of the full origin-destination format, but without the need for huge numbers of cells. With 1,000 statuses and no other dimensions, for instance, the pool format requires only 2,000 cells.

	1980–2000				
	0–29			**30+**	
	Outward	Inward		Outward	Inward
West	2	2	**West**	3	1
East	2	2	**East**	1	3

FIGURE 4.5: Internal migration counts in pool format. The counts are calculated from Panel (c) of Figure 4.4.

A second alternative to the full origin-destination format is to work entirely with net flows, i.e., with the difference between inward movements and outward movements. As can be verified from Figures 4.5 and 4.6, net flows can be calculated by subtracting the 'Outward' column of an array in pool format from the 'Inward' column. Net flows must sum to zero, for each combination of the other dimensions. For instance, in Figure 4.6, net flows for 0–29 year olds are $0 + 0 = 0$, and net flows for 30+ year olds are $-2 + 2 = 0$.

The net format is the most efficient way of measuring the impact of movements on population size, using only half as many cells as the pool format.

1980–2000		
	0–29	30+
West	0	-2
East	0	2

FIGURE 4.6: Internal migration counts in net format. The counts are calculated from Figure 4.5.

However, it is much less amenable to modeling than the pool or origin-destination formats. Net flows are typically small compared to inflows and outflows, which means that small percentage changes in inflows and outflows can produce large percentage changes in net flows.

Another possibility for describing changing statuses is to use 'transitions'. With transitions, the units of measurement are people rather than events. Figure 4.7 illustrates the transitions approach. It shows transitions for the four individuals who were in our hypothetical population in 1980 and in 2010, which can be derived from Panel (a) of Figure 4.1. Two individuals were in West at the start and end of 1980–2000, one was in East at the start and end, and one was in West at the start and in East at the end.

	Region	
Name	**1980**	**2010**
Anna	East	East
Bella	West	West
Cindy	West	East
Doha	West	West

(a) Raw data

Period	**Origin**	**Destination**
1980–2000	East	East
1980–2000	West	West
1980–2000	West	East
1980–2000	West	West

(b) Processed data

1980–2000		
	Destination	
	West	**East**
Origin **West**	2	1
East	0	1

(c) Array

FIGURE 4.7: An array of transitions. The transitions refer to members of the population of 12 fictitious individuals who were alive and in the country in 1980 and in 2010.

Counts of transitions capture different aspects of migration from counts of movements. The one individual in the East-East cell in Panel (c) is Anna. As we saw in Figure 3.2, Anna started and ended the period 1980–2000 in East, but moved into and out of West during the intervening years. The transitions format discards information about the volume of movements made by Anna, but preserves information about beginning and end points.

4.5   Non-Demographic Events

Figure 4.8 shows the construction of an array of tax payments. The process is much the same as the construction of an array of deaths, except that, rather than counting the number of tax payments, we sum the amounts. For instance, rather than simply recording that there were two tax payments by 30+ year olds in West region, we record that the total tax paid was $90+$60 = $150. With non-demographic events, we typically obtain an array of totals, in contrast to demographic events, where we obtain an array of counts.

Name	Year	Age	Region	Amount
Bella	1998	28	West	$50
Cindy	2002	54	East	$80
Emma	1986	41	West	$90
Emma	1995	50	West	$60

(a) Raw data

Period	Age	Region	Amount
1980–2000	0–29	West	$50
1980–2000	30+	East	$80
1980–2000	30+	West	$90
1980–2000	30+	West	$60

(b) Processed data

1980–2000		
	West	**East**
0–29	$50	0
30+	$150	$80

(c) Array

FIGURE 4.8: Constructing an array of tax payments.

4.6 Exposure

Suppose we want to measure propensity of 0–29 year olds to live in East Region in 2010. The choice of numerator and denominator seems fairly clear. The numerator should be the number of 0–29 year olds living in East in 2010 (which is 2), and the denominator should the total number of 0–29 year olds in 2010 (which is $1 + 2 = 3$). (The numbers are obtained from Panel (c) of Figure 4.1.)

Now suppose we want to measure the propensity for 0–29 year olds to die over the period 1980–2000. Instead of measuring the propensity to belong to a particular group, we are measuring the propensity to experience an event. The choice of numerator again seems clear: the number of deaths of 0–29 year olds. But what should we use for the denominator?

One possibility is to use the number of 0–29 year olds at the start of the period. But this number fails to take account of the entry of Grace, Hong, Isabel, Jana, and Kristina into the population of interest, and the fact that Anna and Bella were no longer 0–29 by the end of the period. Another possibility would be to use the number of 0–29 year olds at the end of the period, but this would be subject to similar objections.

The denominator that demographers in fact use when measuring the propensity to experience an event is person-years lived. Person-years lived is ideally calculated by measuring how much each person spends in the population of interest, and adding up. If, for instance, 5 people spend the whole year in the population of interest and 2 people spend half a year, then the total number of person-years lived is $5 \times 1 + 2 \times 0.5 = 6$.

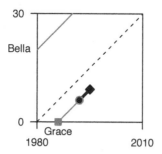

FIGURE 4.9: Calculating exposures for Bella and Grace. The square represents Grace's birth, the circle represents her move from West to East, and the diamond represents her death.

Figure 4.9 shows person-years contributed by Grace and Bella to the age-group 0–29. Between 1986, when she is born, and 1995, when she dies, Grace contributes 9 person-years. Between 1980, when she is 20, and 1990, when she

1980–2000		
	West	**East**
0–29	44	38
30+	83	38

(a) Exact

1980–2000		
	West	**East**
0–29	30	45
30+	90	45

(b) Approximate

FIGURE 4.10: Exact and approximate exposures for the 12 fictitious individuals. The true exposures are calculated from Lexis diagrams in Figures 3.3. The approximate exposures are calculated from the population counts in Figure 4.1.

turns 30, Bella contributes 10 person-years. By carrying out similar calculations for all 12 individuals from Section 3.4, and distinguishing between time spent in West and time spent in East, we arrive at the exposures shown in the Panel (a) of Figure 4.10.

Using individual-level data with exact ages and times is the ideal way to calculate exposure. In many applications, however, the required data are not available. When this happens, demographers resort to approximations. The standard approximation is

$$\text{exposure} = \frac{\text{initial population} + \text{final population}}{2} \times \text{length of period.}$$

Panel (b) of Figure 4.10 shows the results of applying the approximate formula to population counts from Figure 4.1. In this particular example, the approximate values differ substantially from the exact values. The differences are mainly a product of sample size. However, when there are hundreds or thousands of people, rather than just 12, the variation in the individual life histories is averaged away, and the approximations work much better.

4.7   Age, Period, and Cohort

Some events, such as having children or retiring, tend to occur at particular ages. People who were born during the same period tend to experience such events at roughly the same time. People in developed countries who were born in 1950, for instance, generally began to have children in the mid-1970s, and began to retire in the 2010s. This shared timing can lead to shared characteristics. People born during the same period tend, for instance, to have similar family sizes and similar retirement savings. Demographers refer to groups of people born during the same period as cohorts.

The dashed diagonal lines in the Lexis diagrams of Figure 3.3 mark out cohorts for the 12 individuals. The lowest dashed line in each diagram separates people born in 1980–2000 from people born in 1950–1980, for instance, and the next one separates people born in 1950–1980 from people born in 1921–1950.

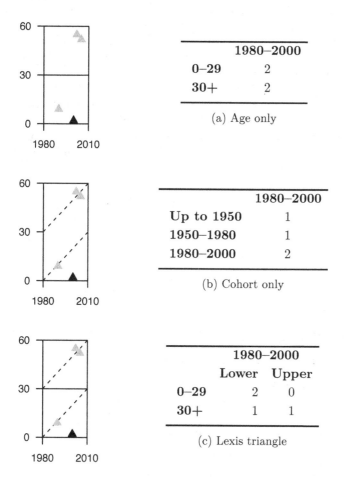

	1980–2000
0–29	2
30+	2

(a) Age only

	1980–2000
Up to 1950	1
1950–1980	1
1980–2000	2

(b) Cohort only

	1980–2000	
	Lower	**Upper**
0–29	2	0
30+	1	1

(c) Lexis triangle

FIGURE 4.11: Arrivals in the country (i.e., immigration) for the 12 individuals, classified by (a) age only, (b) cohort only, and (c) Lexis triangle.

Instead of classifying events by age group, we can classify them by cohort. Figure 4.11 illustrates the alternative classifications, using data on immigration for the 12 individuals. Panel (a) shows an array, and the associated Lexis diagram, that distinguishes age groups but not cohorts. Panel (b) shows an array that distinguishes cohorts but not age groups.

The cohort-based array in Panel (b) cannot be calculated from the age-based array in Panel (a). We cannot, for example, tell whether the two events experienced by age group 0–29 were experienced by the cohort born in 1980–2000, or the cohort born in 1950–1980. Similarly, the age-based array in Panel (a) cannot be calculated from the cohort-based array in Panel (b).

If our array includes one further piece of information, however, we can calculate values for age groups and also for cohorts. The additional piece of information is the 'Lexis triangle' of each event. In Panel (c), the bottom-right triangle is a 'lower' Lexis triangle. Events occurring in this triangle belong to the age group 0–29 and the cohort 1980–2000. The triangle directly above it is an 'upper' Lexis triangle. Events occurring in this triangle belong to the age group 0–29 and the cohort 1950–1980.

The array in Panel (a) can be calculated from Panel (c). For instance, we can add up the counts in the two Lexis triangles for age group 0–29 to obtain the count for age group 0–29. The array in Panel (b) can also be calculated from Panel (c). For instance, we can add up the count for age group 0–29 and upper Lexis triangle, and the count for age group 30+ and lower Lexis triangle, to obtain the count for the cohort 1950–1980.

The relationship between age, cohort, and Lexis triangle, in summary, is:

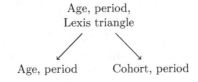

4.8   Rates, Proportions, Means, and Ratios

Demographers calculate many different measures from arrays of counts and totals, but four types of measures are particularly common: rates, proportions, means, and ratios. The measures are summarized in Table 4.1. The 'Min' and 'Max' columns give the minimum and maximum possible values.

TABLE 4.1

Typical definitions of rates, proportions, means, and ratios in demography

Measure	Numerator	Denominator	Min	Max
Rate	Count of events	Exposure	0	-
Proportion	Count, total	Subset of count, total	0	1
Mean	Total	Exposure	-	-
Ratio	Count, total	Count, total	0	-

Often demographers use the term 'rate' to refer to any demographic measure, including proportions, means, and ratios. But when they are being more careful, demographers define a rate as a count of events divided by a measure of exposure. Figure 4.12 illustrates the calculation of rates, using data from Section 3.4 on the population of 12 individuals. The events are births disaggregated by age and region. The exposures are the values from Panel (a) of Figure 4.10.

1980–2000	West	East
0–29	1	0
30+	1	1

(a) Births

1980–2000	West	East
0–29	44	38
30+	83	38

(b) Exposure

1980–2000	West	East
0–29	0.023	0
30+	0.012	0.026

(c) Birth rates

FIGURE 4.12: Calculation of birth rates. Birth rates (c) equal birth counts (a) divided by exposure (b).

Rates can be zero, which happens when the count of events is zero. Rates cannot, however, go below zero. Rates have no definite upper bound, and can go above 1. Consider, for example, a population consisting of one person who dies half way though the year. The death rate in this population is $1/0.5 = 2$ deaths per person-year.

	1980	2010
West	4	4
East	2	4

(a) Original population counts

1980	2010
4	4

(b) Population in West

1980	2010
6	8

(c) Population totals

1980	2010
0.67	0.50

(d) Proportion in West

FIGURE 4.13: Calculating the proportion of the population in West. The numerator (b) and denominator (c) are both obtained from the original set of population counts (a).

With a proportion, the numerator refers to a subset of the people or events in the denominator. The proportion of the year's deaths that occur during January, for instance, is calculated by dividing the count of deaths occurring during January by the count of all deaths for the entire year.

Figure 4.13 shows the proportion of people in West Region in 1980 and 2010. The numerator for the calculations is the top row of Panel (a), and the denominator is the top row of Panel (a) added to the bottom row of Panel (a).

A proportion has an upper limit of 1. A value of 1 occurs when every person or event in the denominator is also included in the numerator.

1980–2000	West	East
0–29	$50	0
30+	$150	$80

(a) Total payments

1980–2000	West	East
0–29	44	38
30+	83	38

(b) Exposure

1980–2000	West	East
0–29	$1.14	$0.00
30+	$1.81	$2.11

(c) Mean payments

FIGURE 4.14: Calculating mean tax payments. Mean tax payments (c) equal total payments (a) divided by exposure (b).

In the models of this book, arrays of means usually describe non-demographic events. We calculate means by dividing totals by exposures. Figure 4.14 shows an example. In Figure 4.14 , the values are all positive, but this is not always the case. In contrast to rates, proportions, and ratios, means can be negative.

1980–2000	West	East
0–29	1	1
30+	2	0

(a) Immigrations

1980–2000	West	East
0–29	1	0
30+	1	1

(b) Emigrations

1980–2000	West	East
0–29	1	-
30+	2	0

(c) Ratio

FIGURE 4.15: Calculating the ratio of immigrations to emigrations. The ratio (c) of immigration to emigration equals immigrations (a) divided by emigrations (b).

A ratio in demography is one set of counts or totals divided by another. Logically speaking, a rate is a type of ratio, but most demographers avoid using the term ratio for a rate. A common problem with ratios is that the value is undefined when the denominator is zero, as happens in Figure 4.15.

4.9 Super-Population and Finite-Population Quantities

Consider a hypothetical population of 10 people in which, during a particular year, no one dies. If we calculate the death rate for the year using observed deaths divided by exposure, then we obtain a death rate of zero. This works well as a description of what actually happened. It does not, however, seem like a good measure of the underlying mortality risks faced by members of the population. It does not seem sensible to claim that the underlying risk was exactly zero.

When distinguishing between underlying propensities and observed values, statisticians refer to 'super-population' quantities and 'finite-population' quantities. Super-population quantities measure underlying risks or propensities, and are always unknown. In the hypothetical population of 10 people, the underlying risk of dying is a super-population quantity. Finite-population quantities measure what actually happened. The death rate of zero in our population of 10 is a finite-population quantity.

Finite-population quantities differ from their super-population equivalents because of the randomness of events such as births and deaths. If, for instance, the super-population risk of dying is 0.1, we would not expect exactly 1 person from a population of 10 to die each year. Instead, we would expect to see year-to-year variation around a long-term average of 1. (For simplicity, we assume here that deaths occur at the end of each year, so that exposure equals population size at the beginning of the year.)

Scientists are generally interested in super-population quantities. They want to measure underlying processes and relationships. Administrators, in contrast, are generally interested in finite-population quantities. They want to know how many people are actually employed, or how many babies were actually born.

A third piece of terminology that often comes up in the context of super-population and finite-population quantities is 'direct estimates'. A direct estimate involves minimal modeling assumptions. The usual direct estimate of a rate, for instance, is the observed number of events divided by exposure.

If the relevant counts or totals are measured without error, then a direct estimate will exactly equal the associated finite-population quantity. If deaths and exposure are measured without error, for instance, then the direct estimate of the death rate will be identical to the finite-population death rate. If, however, the counts or totals contain measurement errors, then the direct estimate and finite-population quantity will differ.

Figure 4.16 summarizes the relationship between super-population quantities, finite-population quantities, and direct estimates. When sample sizes are large, random error tends to be small, relative to the quantities being measured, and super-population and finite-population quantities are numerically similar.

Versions of the super-population versus finite-population distinction appear in demographic textbooks. Many authors distinguish, for instance, between an observed mortality rate for age group x, denoted M_x, and the underlying mortality rate m_x that is used to calculate life expectancy. Demographic textbooks typically use M_x to estimate m_x, however, and almost never use the statistical terms 'super-population' and 'finite-population'.

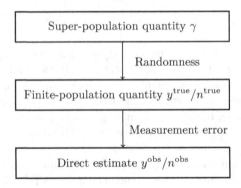

FIGURE 4.16: The relationship between super-population quantities, finite-population quantities, and direct estimates. y^{true} and n^{true} are true values, and y^{obs} and n^{obs} are observed values.

4.10   Collapsing Dimensions

If we were using the list of individuals in Panel (a) of Figure 4.17 to construct an array of population counts disaggregated by age and region, we would add up the number of people in each combination of age and region, to arrive at the array in Panel (b). If, however, we only wanted our array to be disaggregated by age, then we would ignore 'Region' information in the list of individuals, and simply add up the number of people in each age group, to arrive at the array in Panel (c).

Person	Age	Region
Anna	0–29	East
Bella	0–29	West
Cindy	30+	West
Doha	30+	West
Emma	30+	West
Fatima	30+	East

(a) Individuals

	West	East
0–29	1	1
30+	3	1

(b) Population with region

0–29	2
30+	4

(c)   Population without region

FIGURE 4.17: Three views of the population of 12 individuals from Section 3.4 in 1980: the individual-level data, an array that includes region, and an array that does not include region.

If we wanted an array disaggregated only by age, but instead of the original list of individuals, we only had the array in Panel (b), we would follow a superficially different procedure. To obtain the number of 0–29 year olds, we would add the number of 0–29 year olds in West to the number of 0–29 year olds in East, to obtain a total of $1 + 1 = 2$, and similarly for 30+ year olds.

The acts of (i) ignoring a column when aggregating a list of individual people or events, and (ii) adding up across a dimension in an array of counts are fundamentally the same thing. In both cases we are constructing counts that omit one potential classifying dimension. This type of operation is something we do repeatedly in the book, so we give it a specific name: 'collapsing' a dimension.

	West	East
0–29	$1.14	$0.00
30+	$1.81	$2.11

(a) Mean payments

	West	East
0–29	44	38
30+	83	38

(b) Person-years

	West	East
0–29	$50	0
30+	$150	$80

(c) Total taxes

West	East
$200	$80

(d) Total taxes, age dimension collapsed

West	East
127	76

(e) Person-years, age dimension collapsed

West	East
$1.57	$1.05

(f) Annual taxes per person, age dimension collapsed

FIGURE 4.18: Collapsing the age dimension in an array of annual taxes per person.

In most cases, it does *not* make sense to add up a dimension in an array that is not composed of counts or totals, such as an array of rates, proportions, or means. We can nevertheless collapse dimensions in this sort of array. To do so, we have to convert the array to an array of counts or totals, by applying the appropriate weights. Having converted to an array of counts or totals, we collapse dimensions, and then convert back to an array of rates, proportions, or means.

Figure 4.18 illustrates the steps involved in collapsing the dimension of an array of rates, proportions, or means. We have an array of estimated annual taxes per person classified by region and age. We would like to collapse the age dimension.

To do the collapsing, we need an array of person-years to use as weights. We use the exact exposures taken from Panel (a) of Figure 4.10. Multiplying the array of taxes per person by the person-years gives total taxes. We collapse the age dimension by adding up the total taxes in age groups 0–29 and 30+. The array of person-years can be collapsed in the same way. Dividing the collapsed array of total taxes by the collapsed array of person-years yields the collapsed array of taxes per person.

4.11   References and Further Reading

Data on the 12 individuals, plus code to produce the arrays in the chapter, can be found on the website for the book, www.bdef-book.com.

Rees (1985) and Willekens (2006) discuss the measurement of age and time. Schoen (1988) and Rogers (1995) discuss the modeling of movements between statuses, with Rogers emphasizing geographical applications, and Schoen emphasizing sociological ones. Rogers (1990) is a trenchant critic of the concept of net migration. Smith and Swanson (1998) argue that, while inelegant, net migration is nevertheless useful in population modeling. Wilson and Bell (2004) and Alho and Spencer (2005) discuss the pool format. Wachter (2014) is a textbook of demographic methods that emphasizes a cohort perspective on measurements and model.

5

Demographic Accounts

In many demographic applications we focus on a single demographic series, such as births or immigration. But sometimes we need to consider several series simultaneously. To forecast future population counts, for instance, we would normally also forecast future counts for births, deaths, and migration.

It can be helpful, when dealing with multiple demographic series, to treat the series as forming a demographic system. Once we have specified the demographic system, we can describe it using a demographic account.

In this chapter, we introduce demographic systems and demographic accounts, and illustrate the ideas using the population of 12 individuals from Section 3.4. A mathematical description of demographic accounts is given in an optional section at the end of the chapter.

5.1 Demographic Systems

To construct a demographic model, we need to decide whom we want to include in the model, and what sort of characteristics and events we are interested in. When we make these sorts of decisions we are, in effect, defining a demographic system. In applications involving multiple series, it can be helpful to spell the system out explicitly.

The one essential ingredient of a demographic system is a set of **membership criteria** that mark out the population of interest. If a demographic system refers to a national population, for instance, then the population of interest might be everyone who usually lives in the country. If the demographic system is a company, then the population of interest might be everyone employed by the company.

In most cases, a demographic system also involves some sort of **classification scheme** distinguishing between different groups within the population of interest. Some of the characteristics used to define groups may be changeable, such as age or region of residence, while others may be fixed over a person's lifetime, such as region of birth.

A demographic system also usually includes ways of **entering or exiting** the system. The classic ways of entering and exiting a demographic system are births and deaths, but other possibilities include immigrating, emigrating, or attaining a certain age such as 18 or 65.

If some of the characteristics used to define groups are changeable, then the demographic system must recognize **movements** between statuses. If a classification includes geographical region, for instance, and if people are geographically mobile, then a method is needed for measuring migration. If a classification includes age group, then we need to recognize aging as a special type of movement between statuses, in that, provided people remain alive, the movements are completely predictable.

The classic example of a demographic system is a national population. However, the idea of a demographic system has much wider application.

Example 5.1. We can treat the population of the world as a demographic system that includes every human as a member, and classify the population by age, sex, and country of residence. The only way to enter the population is through birth, and the only way to leave is through death. People change their age and country of residence. □

Example 5.2. The population of Aboriginal and Torres Strait Islanders, the indigenous people of Australia, can be modeled as a demographic system. Under the Australian Bureau of Statistics 'self-identification' definition of indigenous status, everyone living in Australia who defines himself or herself as indigenous belongs to the system. Individuals enter the system through birth and immigration, but also by defining themselves as indigenous when they did not previously do so. Individuals exit the system through death and emigration, and by ceasing to define themselves as indigenous. The model of the indigenous population can include all the standard demographic dimensions such as age, sex, and region. □

Example 5.3. A school system can be treated as a demographic system. The members are children on the school roll. Children can be distinguished by age, sex, and year of schooling. Children enter the system by enrolling, and exit by completing their education, dropping out, or dying. Over time, the children age and make their way through the educational hierarchy. □

5.2 Demographic Accounts

A demographic account is a collection of linked demographic arrays describing a demographic system. A demographic account includes an array of population counts, plus one or more arrays of events, which record how people enter, exit, or move within the system.

Every subpopulation within a demographic account must conform to a fundamental accounting identity:

Population		population		count of		count of
count at end	=	count at start	+	entries during	-	exits during
of period		of period		period		period.

The subpopulation could be a single combination of categories, such as females aged 15–19 living in a certain region, or it could be a broader group, such as everyone aged 15–19, or even everyone in the system. Entries and exits are interpreted broadly to include movements into or out of the system, and within the system. If the account includes some form of net flows, then 'entries' can mean 'net entries'. With the exception of net counts, all counts in an account must be non-negative.

The best way to understand demographic accounts is to see some examples. We begin with a very simple one.

5.3   Account with No Region and No Age

We begin with an account, based on the 12 individuals from Section 3.4, that makes no distinction between people within the population. The demographic system underlying the account is summarized in Table 5.1.

TABLE 5.1
Demographic system with no region and no age

Membership	Resident of country
Classification	[None]
Entries	Births, Immigration
Exits	Deaths, Emigration
Movements	[None]

The account is shown in Figure 5.1. The counts for population refer to the years 1980 and 2010, while the counts for births, deaths, immigration, and emigration refer to the period 1980–2000.

The internal migrations from Figure 4.4 do not appear in the account, since the associated demographic system does not include regions. The tax payments from Figure 4.8 also do not appear in the account, since they are non-demographic events, and have no direct effect on population size or structure.

	1980–2000
Births	3
Deaths	2
Immigration	4
Emigration	3

1980	2010
6	8

(a) Population

(b) Components

FIGURE 5.1: A demographic account for the 12 individuals, with no age and no region.

The elements of the account conform to the simple accounting identity:

$$
\begin{aligned}
&\text{population in 2010} &8 \\
=\ &\text{population in 1980} &=6 \\
+\ &\text{births} &+3 \\
+\ &\text{immigration} &+4 \\
-\ &\text{deaths} &-2 \\
-\ &\text{emigration} &-3
\end{aligned}
$$

5.4   Account with Region and No Age

Next we consider a demographic account that distinguishes between regions but does not distinguish between age groups. The underlying demographic system is summarized in Table 5.2. Because we are now distinguishing between regions, our system includes internal migration between regions.

TABLE 5.2
Demographic system with region and no age

Membership	Resident of country
Classification	Region
Entries	Births, Immigration
Exits	Deaths, Emigration
Movements	Internal migration

The account is shown in Figure 5.2. Population, as well as all entries and exits, are now disaggregated by region.

The array describing internal migration in Figure 5.2 is in origin-destination format. We could instead have used an array in pool or net format. Either of the arrays shown in Figure 5.3 could be slotted into Figure 5.2, in place of the existing internal migration array. The accounting identities depend only on net entries, which are the same for all three formats.

	1980	2010
West	4	4
East	2	4

(a) Population

	1980–2000
West	2
East	1

(b) Births

	1980–2000
West	1
East	1

(c) Deaths

	1980–2000
West	3
East	1

(d) Immigration

	1980–2000
West	2
East	1

(e) Emigration

1980–2000	West	East
West	0	5
East	3	0

(f) Internal migration

FIGURE 5.2: A demographic account for the 12 individuals, showing region but not age.

1980–2000	Outward	Inward
West	5	3
East	3	5

(a) Pool format

	1980–2000
West	-2
East	2

(b) Net format

FIGURE 5.3: Alternative formats for the internal migration array in Figure 5.2.

The account of Figure 5.2 is subject to two accounting identities: one for West region, and one for East. The accounting identity for West region is

population of West in 2010	4
= population of West in 1980	= 4
+ births in West	+2
+ immigration to West	+3
+ internal migration from East to West	+3
- deaths in West	−1
- emigration from West	−2
- internal migration from West to East	−5

Net entries due to internal migration are $3 - 5 = -2$. The identity for East region has a similar structure, though for East, net entries due to internal migration are $5 - 3 = 2$, the mirror image of those for West.

5.5   Account with Age and No Region

Finally, we consider a demographic account that recognizes age but not region.
The demographic system is summarized in Table 5.3. Since we are no longer
recognizing differences between regions, internal migration is absent from the
system.

TABLE 5.3

Demographic system with age and no
region

Membership	Resident of country
Classification	Age
Entries	Births, Immigration
Exits	Deaths, Emigration
Movements	Aging

The account is shown in Figure 5.4. To show the accounting indentities
more clearly, we switch from two age groups to three. The account includes not
just age groups and periods, but also Lexis triangles. As we saw in Section 4.7,
including Lexis triangles allows us to work with age groups and with cohorts.

The ability to work with age groups is important because this is how most
users of demographic estimates would like the estimates to be presented. Most
users want death counts, for instance, to be disaggregated by age group, not
by cohort.

The ability to work with cohorts is important because demographic accounting identities are based on cohorts. The accounting identity for the cohort born during 1980–2000 is depicted in Panel (a) of Figure 5.5. The population of 0–29 year olds in 2010 equals the number of births during 1980–2000,
adjusted for immigration, deaths, and emigration occurring within the lower
Lexis triangle for age 0–29. The accounting identity is

$$
\begin{aligned}
&\text{population aged 0–29 in 2010} && 3 \\
&= \text{births during 1980–2000} && = 3 \\
&+ \text{immigration for age 0–29 in lower triangle} && +2 \\
&- \text{deaths for age 0–29 in lower triangle} && -1 \\
&- \text{emigration for age 0–29 in lower triangle} && -1
\end{aligned}
$$

Births in Figure 5.4 are disaggregated by age and Lexis triangle. The ages
and Lexis triangles refer to the mother, not the child. All 3 children are aged
precisely 0 at the time of their birth, and belong to a lower Lexis triangle.

The accounting identity for the cohort born during 1950–1980 is depicted
in Panel (b) of Figure 5.5. With this cohort, which was already alive at the
start of the period, we need to adjust for events in two Lexis triangles, located

	1980	2010
0–29	2	3
30-59	3	3
60+	1	2

(a) Population

	1980–2000	
	Lower	**Upper**
0–29	1	0
30-59	1	1

(b) Births

	1980–2000	
	Lower	**Upper**
0–29	1	0
30-59	0	0
60+	0	1

(c) Deaths

	1980–2000	
	Lower	**Upper**
0–29	2	0
30-59	1	1
60+	0	0

(d) Immigration

	1980–2000	
	Lower	**Upper**
0–29	1	0
30-59	0	1
60+	1	0

(e) Emigration

FIGURE 5.4: A demographic account for the 12 individuals, showing age but not region. "Lower" and "Upper" refer to lower and upper Lexis triangles.

in two different age groups. The accounting identity is

$$
\begin{array}{rl}
\text{population aged 30–59 in 2010} & 3 \\
= \text{population aged 0–29 in 1980} & = 2 \\
+ \text{immigration for age 0–29 in upper triangle} & +0 \\
- \text{deaths for age 0–29 in upper triangle} & -0 \\
- \text{emigration for age 0–29 in upper triangle} & -0 \\
+ \text{immigration for age 30–59 in lower triangle} & +1 \\
- \text{deaths for age 30–59 in lower triangle} & -0 \\
- \text{emigration for age 30–59 in lower triangle} & -0
\end{array}
$$

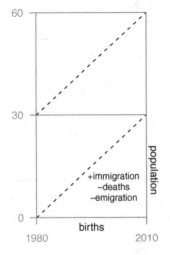

(a) Cohort born in 1980–2000

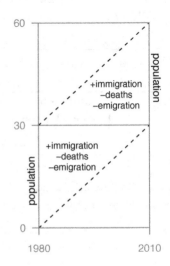

(b) Cohort born in 1950–1980

FIGURE 5.5: Accounting identities based on cohorts.

The accounting identity for the cohort born in 1980–2000 refers to the 3 births that occurred during the period as a single group, without distinguishing the ages or Lexis triangles of the mothers. From the purposes of demographic accounting, only the characteristics of the babies matter; the characteristics of the mothers are irrelevant.

Even though the information is not used in the demographic accounting identities, including the ages and Lexis triangles of the mothers in a demographic account is still valuable in most applications. Users of demographic accounts are often interested in this information on mothers. They may, for instance, want to know if the proportion of births to older mothers is rising over time. Having information on mothers is also useful for statistical modeling, such as modeling age-profiles for fertility rates.

5.6 Movements Accounts and Transitions Accounts*

Starred sections such as this one and the next one contain optional material that can be safely skipped.

The accounts that we have looked at so far in this chapter are known as movements accounts. They relate changes in population to events such as births, deaths, and migration. Demographers have developed a second type of account known as transitions accounts.

Figure 5.6 shows a transitions account based on the 12 individuals. Most full-scale transitions accounts, like most full-scale movements accounts, include age. However, for simplicity, the account in Figure 5.6 only distinguishes between regions.

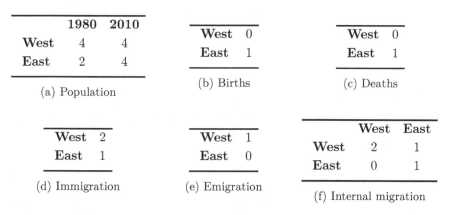

	1980	2010
West	4	4
East	2	4

(a) Population

West	0
East	1

(b) Births

West	0
East	1

(c) Deaths

West	2
East	1

(d) Immigration

West	1
East	0

(e) Emigration

	West	East
West	2	1
East	0	1

(f) Internal migration

FIGURE 5.6: Transitions account showing region but not age.

The population array in a transitions account is identical to the population array in a movements account. The other arrays, however, all refer to transitions (as defined in Section 4.4), rather than events. The births array includes people who were born during the period and were in the system at the end of the period. The immigration array includes people who were not in the system at the beginning of the period, immigrated during the period, and were in the system at the end of the period. The deaths array includes people who were in the system at the beginning of the period and died during the period. The emigration array includes people who were in the system at the beginning of the period, emigrated during the period, and were not in the system at the end of the period. The internal migration array includes people who were in the system at the beginning and end of the period. For all arrays, the regional information refers to the region at the beginning or end of the period.

People who were neither in the system at the beginning of the period nor in the system at the end do not appear anywhere in the account. For instance,

Grace, who was born during the period but who died before the end, does not appear in the account.

Demographic accounting identities for transitions accounts refer to people and transitions rather than people and events, but otherwise have the same structure as the demographic accounting identities for movements accounts. The identity for West Region, for instance, is

$$
\begin{aligned}
&\text{population in West in 2010} && 4 \\
&= \text{population in West in 1980} && = 4 \\
&+ \text{people born during 1980--2000 and in West in 2010} && +0 \\
&+ \text{people who immigrated during 1980--2000 and in West in 2010} && +2 \\
&+ \text{people in East in 1980 and in West in 2010} && +0 \\
&- \text{people in West in 1980 and died during 1980--2000} && -0 \\
&- \text{people in West in 1980 and emigrated during 1980--2000} && -1 \\
&- \text{people in West in 1980 and in East in 2010} && -1
\end{aligned}
$$

5.7   Mathematical Description of Accounting Identities*

In this section, we state the demographic accounting identities for movements accounts in a more detailed form. No existing system of demographic notation quite suits our purposes. For instance, standard notations for multistate life tables assume that there is always an age dimension. Standard notations for multiregional population models become awkward when we do not want to specify the number of classifying dimensions in advance. We therefore introduce our own notation.

Let N denote a demographic array containing population counts. Let B denote a demographic array containing counts of births. Array B may include dimensions describing the child, such as sex of the child, and may also include dimensions describing the parents, such as the age or Lexis triangle of the mother. Counts of children affect future population size, but counts of parents do not. Therefore, we collapse any dimension that does not describe the child, and obtain a collapsed births array B^{ch}. (The collapsing of dimensions is discussed in Section 4.10.) Let C_k $(k = 1, \cdots, K)$ denote demographic arrays containing counts of entries into the system other than births, or counts of exits from the system. Let d_k be an indicator variable that equals 1 if C_k contains entries (including net entries) and -1 otherwise. Let M denote a demographic array containing counts of internal movements. Let M^{net} denote M converted to net format. (Net format is discussed in Section 4.4.)

We need to be able to pick out an individual element of an array without making any assumptions about the number of dimensions that an array contains. We start with a classification system that does not involve age. Let

i denote a unique combination of categories from all dimensions other than the time dimension. For instance, if array $\boldsymbol{A}$ has two sexes, 100 regions, and 10 time periods, then i denotes a combination of sex and region, and has $2 \times 100 = 200$ possible values. For simplicity, we assume that time is measured using 1-year intervals. It is straightforward to extend the equations to deal with intervals of other lengths. If $\boldsymbol{A}$ is an array of population counts, then t indexes an exact time; otherwise t refers to the period between the exact times indexed by t and $t + 1$. We denote by $\boldsymbol{A}[i, t]$ the unique element of $\boldsymbol{A}$ picked out by i and t.

Using the notation given so far, we can state in general form the accounting identity for an account with no age dimension,

$$\boldsymbol{N}[i, t+1] = \boldsymbol{N}[i, t] + \boldsymbol{B}^{\mathrm{ch}}[i, t] + \boldsymbol{M}^{\mathrm{net}}[i, t] + \sum_{k=1}^{K} d_k \boldsymbol{C}_k[i, t]. \qquad (5.1)$$

Turning now to accounts with age, assume that time and age are both measured using 1-year intervals. Age group a includes everyone with exact age larger than or equal to a but smaller than $a + 1$, except that the oldest age group $\mathcal{A}$ includes everyone with exact age $\mathcal{A}$ and over. We also redefine i to be a unique combination of categories from all dimensions other than time, age, and Lexis triangle. We denote by $\boldsymbol{A}[i, a, t]$ the unique element of $\boldsymbol{A}$ picked out by i, a, and t. When picking out an element of an array containing entries, exits or movements, we use the notation $\boldsymbol{A}[i, a, l, t]$, where $l \in \{L, U\}$ denotes the lower or upper Lexis triangle associated with age group a and period t.

Let $\boldsymbol{N}^*$ denote an array measuring 'accession'. Accession to exact age a during period t is the number of people attaining age a during that period. For instance, accession to age 65 during the year 2010 is the number of people having their 65th birthday during 2010.

Accession to age 0 during period t equals births during the period,

$$\boldsymbol{N}^*[i, 0, t] = \boldsymbol{B}^{\mathrm{ch}}[i, t]. \qquad (5.2)$$

Accession to exact ages $a = 1, \ldots, \mathcal{A}$ during period t equals the number of people who were in age group $a - 1$ at exact time t, adjusted for events in the upper Lexis triangle for age group $a - 1$ and period t,

$$\boldsymbol{N}^*[i, a, t] = \boldsymbol{N}[i, a-1, t] + \boldsymbol{M}^{\mathrm{net}}[i, a-1, \mathrm{U}, t] + \sum_{k=1}^{K} d_k \boldsymbol{C}_k[i, a-1, \mathrm{U}, t]. \quad (5.3)$$

Population at exact time $t + 1$, for age groups $a = 0, \ldots, \mathcal{A} - 1$, equals accession to exact age a during the period, adjusted for events in the lower Lexis triangle for age group a and period t,

$$\boldsymbol{N}[i, a, t+1] = \boldsymbol{N}^*[i, a, t] + \boldsymbol{M}^{\mathrm{net}}[i, a, \mathrm{L}, t] + \sum_{k=1}^{K} d_k \boldsymbol{C}_k[i, a, \mathrm{L}, t]. \qquad (5.4)$$

The accounting identity for the oldest age group at exact time $t+1$ involves two groups: people who attain exact age $\mathcal{A}$ during period t, and people who already had already attained age $\mathcal{A}$ at the start of the period,

$$N[i, \mathcal{A}, t + 1] = N^*[i, \mathcal{A}, t] + M^{\text{net}}[i, \mathcal{A}, \text{L}, t] + \sum_{k=1}^{K} d_k C_k[i, \mathcal{A}, \text{L}, t]$$

$$+ N[i, \mathcal{A}, t] + M^{\text{net}}[i, \mathcal{A}, \text{U}, t] + \sum_{k=1}^{K} d_k C_k[i, \mathcal{A}, \text{U}, t]. \quad (5.5)$$

In addition, we have the non-negativity constraints,

$$N^*[i, a, t] \geq 0 \qquad\qquad\qquad (5.6)$$
$$N[i, a, t] \geq 0, \qquad\qquad\qquad (5.7)$$

for all i, a, t.

5.8   References and Further Reading

Zhang (2014) describes the challenges in estimating the size of the indigenous population in Australia.

Rees et al. (2012) and Lomax et al. (2013) are applications of demographic accounts. Rees and Wilson (1977) and Rees and Willekens (1986) discuss the distinction between movements accounts and transitions accounts. Willekens (2006) contains a mathematical description of demographic accounts and related issues.

Preston and Coale (1982) defines accession. To the best of our knowledge, however, demographers have not previously set up demographic accounting identities based on accession and Lexis triangles in the way that we do in Section 5.7.

6

Demographic Data

In the previous two chapters, we looked at the relationship between individual biographies, demographic arrays, and demographic accounts, but without concerning ourselves with sources for the data, or likely errors and gaps. In this chapter, we look briefly at data sources. We review traditional and nontraditional sources of demographic data. We then discuss the modeling choices we have to make when faced with imperfect data.

6.1 Traditional Data Sources

The classic sources of demographic data are vital registration systems—that is, systems for registering births and deaths—and population censuses. Household surveys have generally played a supplementary role, though they can play a central role in very poor countries with few other sources of information.

In rich countries, vital registration systems record virtually all births and deaths, and include basic demographic information such as age of mother and geographic location. However, even the best systems do not generate perfect data. Few countries have foolproof systems for coding addresses, for instance, and some parents take months or years to register the birth of a child. Data on ethnicity are also often incomplete. In most poor countries, vital registration data are unreliable, with a substantial proportion of births and deaths never being registered.

Many countries also have household registration systems, in which people are legally required to notify the government of their place of residence, and any changes in residence. In parts of northern Europe, the system is so comprehensive and efficient that it provides accurate counts of population down to the local level. In most of the rest of the world, household registration systems suffer from problems of under-registration and over-registration. For instance, people do not update their household registration form when they move from their home village to the city.

To obtain counts of population down to the local level, most countries have traditionally relied on population censuses. Censuses try to cover the whole population, and typically do manage to cover over 90% of it. In addition to core demographic variables, such as age and sex, they usually contain additional socioeconomic variables covering topics such as education, occupation, and family status.

For analytical purposes, the most important limitation of censuses is that they occur only once every 5 years, 10 years, or longer. The data become out of date, and are poorly suited to studying short-term trends. Moreover, censuses are highly labor-intensive, which, in many countries, makes them increasingly unaffordable. Statistical agencies have tried to offset rising labor costs through measures such as internet-based collection. However, many countries are developing alternatives to the traditional censuses, and some European countries have already abandoned them.

Household surveys have played a smaller role in demography than they have in other social sciences, because demographers have preferred to use the much larger census and vital registration datasets where possible. However, demographers in countries without censuses or without well-functioning vital registration systems often use surveys to fill the gaps. If traditional censuses become less common in future, household surveys may become more important to demographers, to supply information formerly available through the census.

Household surveys almost always have a complex design, meaning that they use special sampling techniques such as stratification and clustering. With stratification, the target population is divided into groups and different sampling rules are used within each group. With clustering, surveys collect data community-by-community, rather than trying to spread the collection evenly through the whole population. When analyzing data from a complex survey, it is important to take any special features of the design into account.

6.2   New Data Sources

Demographers and national statistical agencies are increasingly turning to data generated as a side product of administrative and commercial processes. Examples include data from tax systems, health systems, electronic payments, mobile phone networks, electricity suppliers, electoral rolls, and social media. These data are opening up new possibilities for demographic modeling. However, administrative and commercial data almost always contain substantial measurement errors, or do not properly cover the population of interest.

For demographic modelers, the greatest obstacle to the routine use of administrative and commercial data is problems with coverage. The population covered by an administrative or commercial dataset rarely matches the population that modelers are interested in. The target population for most demographic modeling is the "usually resident" population, that is, everyone who normally lives in a given region or country. Some administrative systems do, in principle, target the usually resident population. Health systems, for instance, may try to include all residents. However, few healths systems, even in the most orderly and efficient countries, manage to include all the people they are supposed to include, and exclude all the people they are supposed to exclude. For instance, health systems typically miss young people who do not go to the doctor, and fail to remove people who have left the area but not notified the authorities.

Many administrative systems do not even try to target the usually resident population. Tax systems, for instance, deliberately exclude some usual residents, such as people not in the labor force, and include some non-residents, such as taxpayers based overseas. The target populations used by commercial organizations are often very loosely defined. Mobile phone companies, for instance, may not even know how many customers they have because a single billing may cover multiple customers.

Problems of poor or uncertain coverage can often be reduced by linking datasets at the individual level. Advances in technology make it increasingly feasible to identify the same individual in multiple datasets, and to pull together all available information about that individual. The resulting composite datasets can include more people, and more information on each person, than any single dataset. This makes it easier to cover the whole target population, and to determine who should be included and who should be excluded.

Individual-level linking is, nevertheless, more difficult than might be thought. Organizations try to link individuals using names, dates of birth, and addresses, for instance, but people change their names, get their date of birth wrong, and write their address differently on different forms. Moreover, many countries impose legal constraints on data-linking, to protect privacy. And even when records are successfully linked, the data may not provide enough information to dispel all ambiguity about the target population. If a person has not been to the doctor or paid taxes in the past two years, for instance, then the person may have died or left the country, but it is impossible to be sure.

Compounding these problems, administrative and commercial data often do not measure the things that demographic modelers would like them to measure, or they contain measurement errors. The address that the tax system holds for an individual might not be the individual's residential address, but rather the address of the individual's accountant. Police data on ethnicity may describe physical appearance, because that is what is needed to apprehend a person, whereas most social scientists would define ethnicity in terms of cultural affiliation.

But despite all these problems, administrative and commercial data offer unprecedented opportunities to demographic modelers. Administrative and commercial data are often much more up-to-date, and have much more detail on the timing of events, than census or survey data. Whereas a census might ask where a person lived five years ago, administrative and commercial datasets potentially contain every address change a person makes during the period in question. Some administrative and commercial data are more accurate than the census or survey equivalent. Tax data, for instance, provide a far more accurate measure of most people's taxable income than survey data. Administrative and commercial data also contain accurate information on issues that are notoriously difficult to measure using censuses or surveys, such as information on health conditions, criminal offenses, and expenditure on alcohol or junk food.

When datasets contain information on sensitive topics such as health conditions, modelers have extra ethical responsibilities towards respondents, for whom privacy breaches would be especially damaging. This is especially true with administrative or commercial data that were collected for purposes other than data analysis.

An attractive feature of the aggregate-level datasets used in this book is that, compared to individual-level datasets, they raise fewer ethical and privacy issues. When the basic unit is the cell count, rather than the individual record, it is much harder to make discoveries about particular individuals. The fact that aggregate data are less intrusive also means that they are easier to obtain. Organizations are usually more comfortable releasing tabulations with a few basic variables such as age and sex, than releasing individual records. Aggregate data series also generally extend much further into the past than individual-level series, and are found in poor countries as well as rich ones, since they do not require the same sophisticated technologies and processes.

6.3   Data Quality and Model Choice

The arrays in Chapter 4 and the accounts in Chapter 5 all faithfully reproduced the relevant aspects of the lives of the 12 individuals. The number of reported births exactly matched the number of actual births, for instance, and the number of people reported as being alive in 2010 matched the actual number alive in 2010.

The very best demographic data sources, such as death registries in rich countries, approach this level of perfection. But most of the other data that demographers get to work have measurement or coverage errors of varying levels of seriousness.

Researchers and practitioners wishing to apply the methods from this book can choose from the options in Figure 6.1. If there is only one demographic series, they can choose to treat the data as perfect, and select a model from Part III, or choose to treat the data as imperfect, and select a model from Part IV.

If real data are never perfect, why would we ever use one of the models from Part III? With demographic modeling, as with everything else in life, there is never enough time to deal with every problem, so we have to prioritize. In many cases, we can reasonably expect that measurement errors will have much less effect on our results than, say, randomness. If so, it may be better

		Measurement Errors in Data?	
		No	**Yes**
Multiple demographic series	**No**	Part III	Part IV
	Yes	-	Part V

FIGURE 6.1: Models in Part III deal with single demographic series measured without error; models in Part IV deal with single demographic series measured with error; and models in Part V deal with multiple demographic series measured with error.

to spend time understanding and responding to randomness than to spend time on measurement errors.

Figure 6.1 has no entry in the bottom-left corner. In other words, we do not allow for the possibility of multiple demographic series with no measurement errors. The reason is that demographic accounts make it much harder to pretend that the data are perfect. If we put observed population counts, births, deaths, and so on into a demographic account, we will inevitably find that the some elements of the account do not conform to the demographic accounting identities. If the count of reported births overstates the true count by one birth, for instance, then, unless there are exactly offsetting errors in other demographic series, all the associated accounting identities will be off by one.

6.4 References and Further Reading

Siegel (2003) and Smith et al. (2013) discuss demographic data sources, focusing mainly on the United States. Moultrie et al. (2013) discuss data sources and data quality, with an emphasis on developing countries. Their work is available for free at *demographicestimation.iussp.org*. Coleman (2013) reviews international efforts to replace traditional censuses.

Part II

Bayesian Foundations

7

Bayesian Foundations

In Part II, we review the statistical techniques that we will need for estimating and forecasting demographic arrays and accounts.

The statistical methods used in this book belong to a branch of statistics known as Bayesian statistics. In this chapter, we describe the distinctive features of Bayesian statistics. In the next two chapters, we introduce specific concepts and techniques that we need for the models of this book.

7.1 Bayesian Statistics

Most readers of this book who have taken a course in statistics will have learnt 'classical' or 'frequentist' statistics. Bayesian statistics is an alternative approach, whose founding document, "An Essay towards solving a Problem in the Doctrine of Chances" by the Reverend Thomas Bayes, was read to the Royal Society in 1763.

Over most of the 20th century, frequentist methods were more widely taught and used than Bayesian ones. One reason is mathematical tractability. For many standard problems, frequentists methods require less complicated calculations than their Bayesian equivalents. The second reason is a desire for objectivity. Frequentist methods appear to offer a way of drawing conclusions from data alone, with a minimal role for personal judgement.

Since the 1980s, however, the tractability argument for frequentist methods has been losing force. New mathematical techniques for carrying out Bayesian inference have appeared, and computers have taken over from pencils and paper. Moreover, statisticians have been building ever more complex models, and for complex models, Bayesian methods often turn out to be easier, rather than harder, to implement than frequentist ones.

The objectivity argument has perhaps lost force as well, as Bayesians have defended the subjective elements of Bayesian methods, and drawn attention to the subjectivity in frequentist approaches. However, practical advantages—the ability to fit bigger models—have probably done more to boost the popularity of Bayesian methods than philosophical considerations.

7.2 Features of a Bayesian Data Analysis

An idealized Bayesian analysis has three steps:

1. **Model specification.** Specify a joint probabilistic model for known demographic quantities in the data, unknown demographic quantities of interest, as well as other unknown quantities needed in the model.

2. **Inference.** Infer the unknown quantities from the known quantities and the specified model, and express the inferences probabilistically.

3. **Model checking.** Assess the quality of the inferences, by examining the agreement with the data and the substantive implications of the model, and trying alternative specifications.

The frameworks set out in Figures 1.4, 1.5, and 1.6 in Chapter 1 give schematic view of the probabilistic models in this book. Figure 7.1 is an expanded version of Figure 1.5. As well as distinguishing between super-population quantities and finite-population quantities (as defined in Section 4.9), it highlights places where the model uses probabilities.

The most straightforward use of probabilities in Figure 7.1 is for modeling randomness. The true array of finite-population counts is subject to random variation, even when the associated arrays of super-population rates and exposures are fixed. The true number of deaths, for instance, is a random quantity, even if the death rate and the number of person-years exposed to the risk of dying are known with certainty. The contents of the observed datasets are also subject to random variation. For instance, even if the probability of registering a death, and the number of deaths per year, were to remain constant, the number of registered deaths would still vary from year to year, since registration is a random event.

A second use of probabilities is modeling limits to knowledge. Using probabilities to describe limited knowledge is common in everyday life. People say, for instance, that they are "90% sure" about their answer in a general knowledge quiz. Bayesians, unlike frequentists, are willing to use probabilities to describe limited knowledge about uncertain quantities inside a statistical model.

In Figure 7.1, the prior model describing the rates, probabilities, or means reflects limited knowledge. We might, for instance, have some knowledge on how mortality rates vary with age, sex, and time, but we could not be 100% certain about the strength of these relationships. Similarly, the contents of the observed datasets are also subject to limited knowledge. For instance, we might know that each death had a small chance of being missed by the registration system, but we would typically not know exactly how small this chance was.

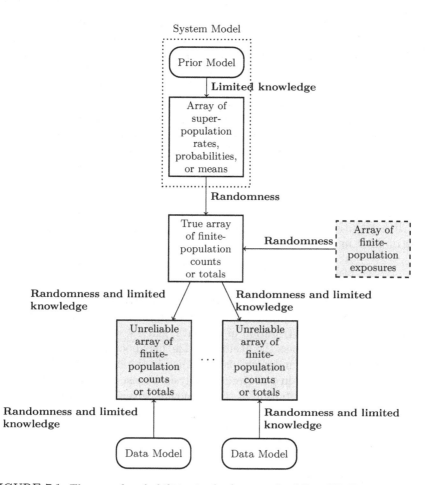

FIGURE 7.1: The use of probabilities in the framework of Part IV. Features modeled using probabilities are shown in **boldface**. Note that, unlike in Figure 1.5 in Chapter 1, the diagram distinguishes between super-population quantities and finite-population quantities.

The term for the probabilities associated with a set of quantities is the 'probability distribution' for the quantities. We will return to the topic of probability distributions in Chapter 8.

As illustrated in Figure 7.1, the probabilistic model for all quantities is constructed piece by piece. The prior model specifies the probability distribution for the super-population rates, probabilities, or means. Given the rates, probabilities, or means, and possibly the exposure, we specify the probability distribution for the true array of finite-population counts or totals. Given the true array of finite-population counts or totals, the data model specifies the probability distribution for the observed unreliable arrays of finite-population counts or totals.

The second step in the analysis process, inferring the unknown quantities from the observed data, also uses probability distributions. In mathematical terms, our probabilistic model for the data and unknown quantities amounts to a giant probability distribution,

$$p(\text{unknowns}, \text{data}),$$

which is read as "the joint probability distribution of the unknowns and data".

It is traditional, in Bayesian analyses, to decompose the joint probability distribution into two terms,

$$p(\text{unknowns}, \text{data}) = p(\text{unknowns})p(\text{data}|\text{unknowns}).$$

The first term, $p(\text{unknowns})$, is the probability distribution of the unknowns, and is referred to as the prior. The second term, $p(\text{data}|\text{unknowns})$, is the conditional probability distribution of the data, given a value for the unknowns. This term is referred to as the likelihood. The likelihood summarizes whatever information about the unknowns is contained in the data to hand.

The inference step in a Bayesian analysis consists of deriving the conditional probability distribution of the unknowns, given the data,

$$p(\text{unknowns}|\text{data}),$$

This distribution is referred to as the 'posterior distribution'. It is the principal output from a Bayesian analysis.

"Prior" and "posterior" are Latin for "before" and "after". A traditional way of defining prior and posterior distributions is that the prior distribution describes the analyst's beliefs before seeing the data, and the posterior distribution describes the analyst's beliefs after seeing the data. These sorts of definitions give a misleading impression of modern Bayesian statistical practice. Most modelers in fact formulate their prior distributions after seeing the data. In addition, most modelers do not try to use prior and posterior distributions to describe their own beliefs, but rather to describe some version of "what it is reasonable to believe, given available information". These points will become clearer in later chapters.

The task of deriving the posterior distribution can be technically demanding. We discuss calculation of the posterior distribution in Section 9.1, but our general approach in this book is to rely on the software to look after the details for us.

Model checking is an essential part of the overall analysis. In the course of specifying a model, we inevitably simplify aspects of the system that we are interested in, and omit other aspects. We need to check that these simplifications and omissions do not substantively affect our results. With model checking, we probe our assumptions, and look for potential problems. If problems are reviewed, we return to step 1 of the analysis process, and reformulate our model.

7.3   References and Further Reading

Courgeau (2012) reviews the history of probability theory, Bayesian and frequentist methods, and their close relationship with the development of the social sciences, including demography.

The three-step description of Bayesian data analysis is set out in Gelman et al. (2014, p.3). Gelman et al. (2014) have been prominent proponents of the idea that model checking deserves the same emphasis within Bayesian analyses as model building and inference.

Gelman et al. (2014) is a standard text on Bayesian methods, which we draw on throughout this book. However, readers who are comfortable with mathematics, but who are new to Bayesian methods, may find Hoff (2009) an easier place to start. Alternatively, McElreath (2016) has less on the mathematics than Hoff (2009), but more on the ideas behind the methods.

For a two-page discussion of uncertainty due to randomness versus uncertainty due to limited knowledge, see O'Hagan (2004). Hájek (2012) discusses the leading interpretations of probability, and identifies problems with all of them.

Arguments for using probability to represent limited knowledge can be made on empirical grounds as well as philosophical ones. Tetlock and Gardner (2015) present evidence from a large study of forecasting performance showing that people who are good at using probabilities to measure uncertainty make more accurate forecasts.

8

Bayesian Model Specification

This chapter discusses ideas and issues related to specifying Bayesian demographic models. We begin with discussion of basic probability distributions. We then explain how complicated Bayesian models can be constructed using the basic probability distributions. Main ideas include exchangeability, hierarchy, and incorporating external information.

8.1 Using Probability Distributions to Quantify Uncertainty

A probability distribution is a way of quantifying or modeling uncertainty. It assigns probabilities to events or to empirical statements.

Example 8.1. Table 8.1 shows a probability distribution in which the event being modeled is the number of deaths occurring during a year in a population of three people. The probability of no deaths occurring is 0.216, the probability of 1 death occurring is 0.432, and so on.

TABLE 8.1
Probability Distribution
for Number of Deaths

Deaths	Probability
0	0.216
1	0.432
2	0.288
3	0.064
Total	1.000

□

Example 8.2. Figure 8.1 depicts a probability distribution quantifying current knowledge about the size of a population. The most likely values for the size of the population are 3 and 4, but they could plausibly be as high as 15.

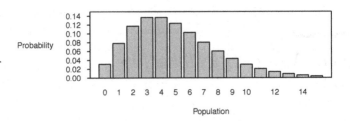

FIGURE 8.1: A probability distribution summarizing beliefs about a population count. □

The probability distribution in Example 8.1 is describing uncertainty due to randomness, and the probability distribution in Example 8.2 is describing uncertainty due to limited knowledge. As we saw in Chapter 7, Bayesians use probability distributions to describe both sorts of uncertainty.

One thing that the probability distributions in Examples 8.1 and 8.2 have in common is that both are 'discrete'. A probability distribution is discrete if the outcomes can be listed in a sequence, and a probability can be assigned to every possible outcome.

Many of the important outcomes in demography, such as counts of events or people, are discrete. Some, however, are not. Examples are body weight, hourly wage, or health expenditure, all of which can, in principle, be measured to an infinite number of decimal places. Mathematicians describe these sorts of variables as 'continuous'.

With a continuous variable, it does not make sense to calculate the probability of a particular value. It does not, for instance, make sense to calculate the probability that a person's body weight is exactly 76.3427478... kilograms. Instead, we calculate probabilities over ranges of values. For instance, we can calculate the probability that a person's weight is between 76.3 kg and 76.4 kg, or the probability that a person's weight is greater than 75.5 kg. We assign a 'probability density' over the range of the variable, and calculate probabilities by aggregating ('integrating') the density.

Example 8.3. Figure 8.2 shows the probability distribution for body weight in a hypothetical population with an average body weight of 75 kg. The probability that a randomly-chosen person from the population has a weight of 75.5 kg or more is equal to the area of the shaded region under the density curve. This area equals 0.048.

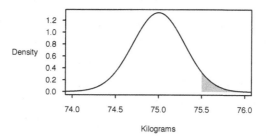

FIGURE 8.2: Probability distribution for body weight in a hypothetical population with a mean body weight of 75 kilograms. □

Example 8.4. Figure 8.3 depicts a continuous probability distribution used to describe limited knowledge on an annual probability of dying. The most likely value is 0.4, but the value could plausibly be as high as 0.6 or as low as 0.2.

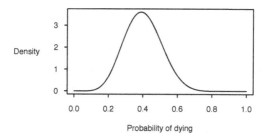

FIGURE 8.3: A distribution summarizing beliefs about an annual probability of dying. □

8.2 Posterior as a Compromise between Likelihood and Prior

As discussed in Section 7.2, a Bayesian analysis involves a likelihood, prior, and posterior:

Likelihood $p(\text{data}|\text{unknowns})$. A function showing the probability of observing the data, for any particular value of the unknowns. A way of summarizing the information about unknowns contained in the data.

Prior distribution $p(\text{unknowns})$. A probability distribution summarizing information about the unknowns, beyond what is contained in the data.

Posterior distribution $p(\mathbf{unknowns|data})$. A probability distribution summarizing information about the unknowns, after combining information from the likelihood and prior.

The posterior distribution is a compromise between the likelihood and the prior. In most cases, the likelihood contains more information about the unknowns than the prior, in the sense that it is concentrated on a narrower range of values than the prior. In such cases, the posterior resembles the likelihood more strongly than the prior. But sometimes the prior contains more information than the likelihood, and the prior dominates.

Example 8.5. Figure 8.4 shows the likelihood, prior, and posterior for the annual probability of dying. The likelihood is more strongly peaked than the prior, implying that it concentrates on a narrower range of values than the prior—i.e, it is more informative than the prior. The posterior, accordingly, looks more like the likelihood than the prior. However, the prior, which favours values near 0.4, does have some effect, pulling the posterior a little away from the likelihood, which favours values near 0.3.

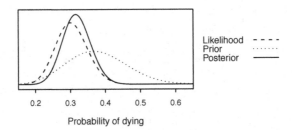

FIGURE 8.4: For combinations of likelihood and prior, and the resulting posteriors, for the annual probability of dying.   □

If the likelihood changes because we alter our modeling assumptions or acquire more data, then the posterior will change. Similarly, if the prior changes, then the posterior will change.

Example 8.5 (continued). Figure 8.5 shows four combinations of likelihood and prior, and the resulting posteriors, for the annual probability of dying. The likelihood and the prior come in two versions: a weak version with a flat peak, and a strong version with a sharp peak. Comparing the two bottom panels with the two top panels, we can see that a stronger prior pulls the posterior further away from the likelihood. Similarly, comparing the two right-hand panels with the two left-hand panels, we can see that a stronger likelihood pulls the posterior further away from the prior.

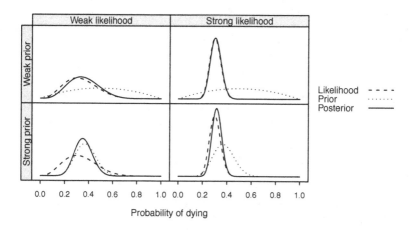

FIGURE 8.5: Four combinations for likelihood, prior, and posterior for the annual probability of dying. □

In general, the more data we have, the stronger our likelihood will be, and the more that it, rather than the prior, will determine the shape of the posterior. This is a fundamental principle of Bayesian statistics.

8.3 Standard Probability Distributions

Mathematicians have developed many standard probability distributions, in which, by providing values for a small number of parameters, we can generate probability for any possible outcome in the discrete case, or over any range of values in the continuous case. Standard probability distributions are the basic building blocks of probabilistic models.

We introduce the Poisson, normal, and binomial distributions below. We also look at the half-*t* distribution, which is less well-known than the other distributions, but which plays an important role in the complicated models that we will be using in Parts III, IV, and V.

8.3.1 Poisson Distribution

The Poisson distribution is used when the outcome in question is some sort of count without definite limit. In its classic form, the Poisson distribution has a single parameter, describing the mean value for the outcome. In demographic modeling, however, we often specify the Poisson distribution so that it includes an exposure term and a rate parameter. (See Section 4.6 for a definition of exposure.) When modeling fertility, for instance, the exposure term is usually person-years lived by reproductive-age women, while the rate parameter is births per person-year lived.

We use

$$y \sim \text{Poisson}(\lambda)$$

to signify that y is drawn from a Poisson distribution with mean λ. We use

$$y \sim \text{Poisson}(\gamma w)$$

to signify that y is drawn from a Poisson distribution with rate γ and exposure w.

In a Poisson distribution in the counts form, the probability that a count equals y is

$$p(y) = \frac{1}{y!} \lambda^y \exp(\lambda),$$

where $\exp(\cdot)$ is an exponential function with base e. In a Poisson distribution in the rates-exposure form, the probability that a count equals y is

$$p(y) = \frac{1}{y!} (\gamma w)^y \exp(\gamma w).$$

Example 8.6. Figure 8.6 shows a Poisson distribution with exposure 10 and rate 0.3. We could, for instance, use this distribution to model the count of births in a population of women that contained an average of 10 people over a one-year period, where the fertility rate was 0.3 births per person-year lived.

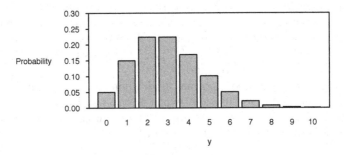

FIGURE 8.6: Poisson distribution with exposure 10 and rate 0.3. The horizontal axis shows the count, and the vertical axis shows the associated probability. □

In this book, we mainly use the Poisson distribution in its rates-exposure form. When modeling counts of people, rather than events, however, the idea of exposure is no longer relevant, and we revert to the classic form.

8.3.2 Binomial Distribution

The binomial distribution is used to model the number of "successes" out of a given number of "trials". It has two parameters, the total number of trials, and the probability that each trial will result in a success. The probability distribution in Example 8.1, for instance, is a binomial distribution with 3 trials and a probability of 0.4.

We use

$$y \sim \text{binomial}(n, \pi)$$

to signify that y is drawn from a binomial distribution, where n is the total number of trials and π is the probability of succeeding.

In the binomial distribution, the probability that there are y successes is

$$p(y) = \binom{n}{y}\pi^y(1 - \pi)^{n-y}, \tag{8.1}$$

where $\binom{n}{y}$ is the number of ways of selecting y elements from a set of n elements, ignoring the selection order.

Example 8.7. Figure 8.7 shows a binomial distribution with 10 trials and probability 0.3. We could, for instance, use this distribution to model diabetes prevalence in a population of 10 people where the probability that each person had diabetes was 0.3.

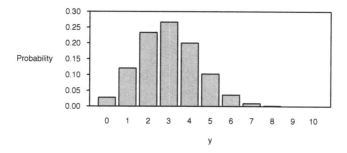

FIGURE 8.7: Binomial distribution with 10 trials and probability 0.3. The horizontal axis shows the count of successes, and the vertical axis shows the associated probability. □

8.3.3 Normal Distribution

The normal distribution is a continuous probability distribution that has an unlimited range, but that places a large probability on central values. The normal distribution has two parameters: the mean and the standard deviation.

The mean specifies the center of the distribution. The standard deviation specifies variability: the larger the standard deviation, the higher probability there is for values a long way from the mean.

We write

$$y \sim N(\mu, \sigma^2)$$

when y is drawn from a normal distribution with mean μ and standard deviation σ. The term σ^2 is called variance. In Example 8.3, the weight of a randomly-chosen person follows a normal distribution with $\mu = 75$ and $\sigma = 0.3$.

The probability density function for a normally-distributed variable y is

$$p(y) = \frac{1}{\sqrt{2\pi}\sigma} \exp\left(-\frac{(y-\mu)^2}{2\sigma^2}\right). \qquad (8.2)$$

An important fact about the normal distribution, which we will use repeatedly in later chapters, is that a quantity modeled by normal distribution has an approximately 95% chance of being within two standard deviations of the mean. (The exact number is closer to 95.4%.)

Example 8.8. Figure 8.8 shows a normal distribution with mean 0 and standard deviation 1. The gray area lies within two standard deviations of the mean. It includes approximately 95% of the total probability of the distribution.

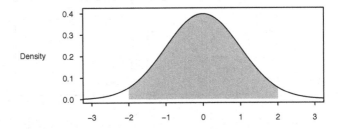

FIGURE 8.8: Normal distribution with mean 0 and standard deviation 1. The horizontal axis shows the value of the variable, and the vertical axis shows the associated probability density. □

8.3.4 Half-*t* Distribution

Our final distribution is not generally used for modeling outcomes directly, but instead is used for modeling standard deviations within a larger model. The half-t distribution prohibits negative values, and favours low positive values over high ones.

We use the notation

$$y \sim t_\nu^+(\sigma^2)$$

when y has a half-t distribution with ν degrees of freedom and scale σ. The scale parameter governs the overall shape of the distribution. Lower values for scale lead to a distribution that is more concentrated around small values. The degrees of freedom parameter governs the size of the tail. Lower values for degrees of freedom lead to a "heavier" tail, implying that there is a greater chance of occasional large values.

The probability density function for a half-t distributed variable y is

$$p(y) = \frac{2\Gamma((\nu+1)/2)}{\Gamma(\nu/2)\sqrt{\nu\pi}\sigma} \left(1 + \frac{1}{\nu}\left(\frac{y}{\sigma}\right)^2\right)^{-(\nu+1)/2},$$

where $\Gamma(\cdot)$ is the Gamma function.

Example 8.9. Figure 8.9 shows a half-t distribution with seven degrees of freedom and scale 1. The distribution is weighted towards values below 1, but does not completely rule out values higher than 3.

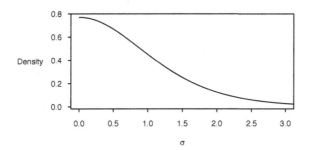

FIGURE 8.9: Half-t distribution with seven degrees of freedom, and scale 1. The horizontal axis shows the value of the variable, and the vertical axis shows the associated probability density. □

Example 8.10. Figure 8.10 shows half-t distributions with seven degrees of freedom and three alternative values for the scale parameter: 1, 0.1, and 0.01. Because the three distributions have such different shapes, it is only possible to show part of each curve within the same graph. It is, nevertheless, clear that a half-t distribution with a scale of 0.01 has a much more restricted range than a half-t distribution with a scale of 1, and that a half-t distribution with a scale of 0.1 occupies an intermediate position.

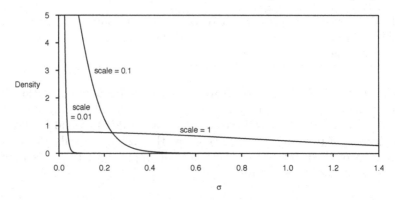

FIGURE 8.10: Half-*t* distributions with seven degrees of freedom, and scales of 1, 0.1, and 0.01.   □

8.4   Exchangeability

When assembling models out of standard probability distributions, statisticians almost always invoke, in some form, the idea of exchangeability. Units in a statistical model are exchangeable if the ordering of the units conveys no information about those units. Randomly swapping the IDs of the units would have no effect on our expectations about values of the units, or the relationship between them.

Example 8.11. A classic example of exchangeability is survey respondents chosen through simple random sampling. If we have a list of the whole target population of N people, and choose each person in the sample by drawing a random number between 1 and N, then respondents are exchangeable. Respondent 100 has precisely the same probability of having a college degree as respondent 1,000. Moreover, learning that respondent 99 has a college degree is no more helpful for deciding whether respondent 100 has a college degree than learning that respondent 1,000 has a college degree.

If, instead of using simple random sampling, we randomly selected households and interviewed everyone in the household, and if household members were listed together in the dataset, then respondents would not be exchangeable. Respondents 99 and 100 would be more likely to belong to the same household than respondents 99 and 1,000. Knowing that respondent 99 had a college degree *would* tell us more about the likely education of respondent 100 than knowing that respondent 1,000 had a college degree.   □

In the models of this book, the basic unit of analysis is not an individual respondent, but a cell within a demographic array. When, in this book, we make assumptions about exchangeability, we are referring to cells within an array, rather than particular individuals or events. We might, for instance, assume that cells representing different regions are exchangeable, or that, for a given combination of age, sex, and education, cells representing different occupations are exchangeable.

Example 8.12. When building a model of fertility rates, we might assume that the underlying rates are exchangeable across regions. This would entail assuming that fertility rates in regions 1 and 2, for example, are no more likely to be similar than fertility rates in regions 1 and 50.

One common strategy, which we will examine in detail in Part III, is to assume that region-specific rates, after a log transformation, are random draws from the same normal distribution. (For a description of the log transformation, see Section 12.2.) The number of events in each region is then modeled as a random draw from a Poisson distribution. Applying this approach to data on births leads to the model depicted in Figure 8.11.

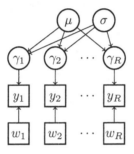

FIGURE 8.11: A model for regional fertility rates, where regions are exchangeable. Region r has y_r births, w_r person-years for reproductive-age women, and underlying fertility rate γ_r. The γ_r are exchangeable, and, after a log transformation, are drawn from a common distribution with parameters μ and σ.

We can write this model as

$$y_r \sim \text{Poisson}(\gamma_r w_r),$$
$$\log \gamma_r \sim \text{N}(\mu, \sigma^2).$$

□

Assuming that units are exchangeable is *not* the same as assuming that they are identical. In Example 8.11, for instance, we do not assume that respondents are identical to one another, and in Example 8.12, we do not assume that regions all have the same underlying fertility rates.

Instead, exchangeability is a way of characterizing our knowledge about a set of units. When we treat units as exchangeable, we are claiming that we know enough about the set that the units belong to, or about the process that generated them, to justify putting them together into a group, and modeling them in the same way. However, we are also claiming that we do not know enough about each particular unit to say which units have low values and which have high values.

8.5   Partial Exchangeability

In demographic estimation and forecasting, where the basic unit of analysis is a cell classified by dimensions such as age, sex, and region, pure exchangeability is rare. When modeling regional variation in mortality, fertility, or migration, for instance, we typically have some idea about which regions are likely to have high values and which are likely to have low values, based on regional income levels or ethnic composition. If the violations of exchangeability are small, then it may be sensible to ignore them, sacrificing realism for convenience. If not, we may need to turn to a more complicated form of exchangeability.

8.5.1   Exchangeability within Groups

Rather than assuming that every unit is exchangeable with every other unit, we may decide to divide the units into groups, and only assume exchangeability within each group.

Example 8.12 (continued). Consider again the problem of modeling regional variation in fertility rates. Urban areas typically have lower fertility rates than rural areas. If some regions are predominantly urban, while others are predominantly rural, then treating all units as exchangeable may not be appropriate. Instead, we might separate the regions into an urban group and a rural group. We would treat urban regions as exchangeable only with other urban regions, and rural regions as exchangeable only with other rural regions.   □

8.5.2   Exchangeable Residuals

More generally, if we have information about the units that we think could help explain outcomes, then we can include that information in the model through the use of covariates. A covariate—also known as an independent variable, explanatory variable, or predictor—is a measurement on each unit that can help predict outcomes for that unit. In a model of regional employment rates, for instance, regional GDP per capita is a possible covariate.

When we include a covariate in a model, we measure the extent to which low or high values for the outcome are consistently associated with low or high values for the covariate. In a model of regional unemployment rates, for instance, we might find that employment tends to be high in regions with high

GDP per capita, and low in regions with low GDP per capita. If so, then we can use a region's GDP per capita to predict the region's employment rate.

In models with covariates, we apply the assumption of exchangeability to the 'unexplained' part of each unit's outcome. We take the observed outcome for each unit and subtract off the outcome that we predict from the covariates. The remainder, which statisticians call the residual or error, is due to factors other than the covariates. We treat the errors, rather than original outcomes, as exchangeable.

Example 8.12 (continued). Figure 8.12 shows a model for regional fertility rates that includes covariates. The fertility rate for each region, after logarithm transformation, is a random draw from a normal distribution, whose mean depends on the region-specific covariates.

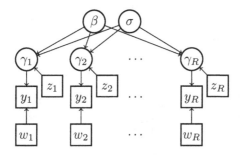

FIGURE 8.12: A model for regional fertility rates in which regions are exchangeable after adjusting for covariates. Region r has y_r births, w_r person-years for reproductive-age women, and underlying fertility rate γ_r. Regional differences in the γ_r can partly be explained by regional differences in covariates z_r. The relationship between the covariates and fertility rate is measured by β.

The revised model can be written as

$$y_r \sim \text{Poisson}(\gamma_r w_r),$$
$$\log \gamma_r \sim \text{N}(z_r \beta, \sigma^2).$$

□

8.5.3 Exchangeable Increments

With some types of units, such as units of age or time, order almost always matters. We would almost always expect fertility rates in 2015 to be more strongly related to fertility rates in 2014 than to rates in 1950. Similarly, we would almost always expect obesity prevalence among 60 year olds to be more strongly related to obesity prevalence among 59 year olds than to prevalence among 10 year olds.

When order matters, treating units as exchangeable is not appropriate. Exchangeability is, nevertheless, an essential ingredient of most models of variation over age or time. Instead of applying exchangeability assumptions to the units themselves, we apply them to differences between adjacent units.

Example 8.13. The sex ratio at birth is the number of male births per 100 female births. Panel (a) in Figure 8.13 shows sex ratios in the United Kingdom from 1938 to 2014. There is a clear shift in the average ratio around 1980, and a hint of a shift around 1940. Knowing the sex ratio in 1954 and 1956 is more helpful for predicting the sex ratio in 1955 than knowing the sex ratio in 1984 and 1986. It does not appear to be appropriate to treat annual sex ratios as exchangeable.

Panel (b) shows the annual *change* in the sex ratio at birth, that is, it shows the rate in 1939 minus the rate in 1938, the rate in 1940 minus the rate in 1939, and so on. In contrast to the sex ratios themselves, the changes do seem to be exchangeable.

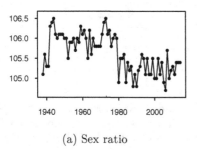

(a) Sex ratio

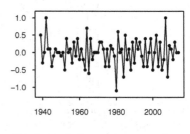

(b) Annual change in sex ratio

FIGURE 8.13: Sex ratio at birth in the United Kingdom.   □

When unit-to-unit differences appear to be exchangeable, though the units themselves are not, a natural model for the units is a random walk. Figure 8.14 depicts a random walk model. The value for the current unit equals the value for the previous unit plus a random quantity. The random quantities are referred to as 'errors' or 'innovations', and are assumed to be exchangeable.

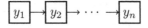

FIGURE 8.14: A random walk model.

A random walk model typically has the form

$$y_i \sim \mathrm{N}(y_{i-1}, \sigma^2).$$

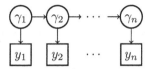

FIGURE 8.15: A local level model. The γ_i represent the expected value for the y_i. The model has two types of errors: ones that affect a γ_i, and hence permanently change the expected value, and ones that only affect a y_i, and hence are transient.

The sex ratio data discussed in Example 8.13 could be modeled using a random walk, given that the year-on-year differences shown in the right-hand panel of Figure 8.13 appear to be exchangeable. This model would, however, miss an important feature of data. It would fail to distinguish permanent shifts, such as the one that occurred around 1980, from annual variation.

A local level model, as depicted in Figure 8.15, distinguishes between permanent shifts and transient noise. A local level model allows for two types of error: (i) errors that permanently change the mean for subsequent values, and (ii) errors that only affect the current value. Both types of error are treated as exchangeable.

A typical local level model has the form

$$y_i \sim N(\gamma_i, \tau^2),$$
$$\gamma_i \sim N(\gamma_{i-1}, \omega^2).$$

8.6 Pooling Information

Imagine that we are in a fruit shop, and have just weighed four apples. The apples' weights were 82, 96, 75, and 89 grams. If the next piece of fruit is an apple, then we might expect it to weigh somewhere between 80 and 100 grams. But if the next piece of fruit is an orange, then we would be much less confident about its weight. The reason is exchangeability. Apples are exchangeable, while apples and oranges are not.

When units are exchangeable, we can pool information from across the units. Knowledge about any one unit carries over, at least partly, to all the other units.

Now imagine that, instead of 82, 96, 75, and 89, the apples' weights were 89, 91, 89, and 90 grams. Based on these observations, if the next piece of fruit was an apple, we might say that its weight was somewhere between 88 and 92 grams, rather than 80 and 100.

Stating that an apple's weight is likely to be between 88 and 92 grams is a much stronger claim that stating that the apple's weight is likely to be between 80 and 100 grams. When variability between units is low, we can pool information more readily, and make stronger inferences.

In summary, the relationship between exchangeability, variability, and the pooling of information is:

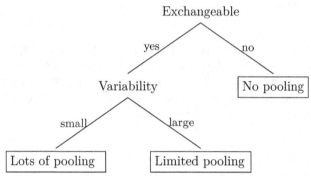

We can build these ideas into our probabilistic models. A model of apple weights can treat the weight of each apple as a draw from a probability distribution. If the weights of the apples were dispersed, e.g. 82, 96, 75, and 89, then the parameter of the probability distribution that governed variability could reflect this. If the weights of the apples were concentrated, e.g. 89, 91, 89, and 90, then the parameter could again reflect this. The variability parameter would determine the amount of information pooling.

Bayesian models typically carry out 'partial pooling' of information. The estimate for each unit is a compromise between the (i) measurement for that particular unit, and (ii) the average measurement across all units that it is exchangeable with.

With the sort of data that we deal with in this book, there is one further twist to the relationship between variability and pooling of information. The basic unit in our data is a cell within a demographic array, rather than an individual person (or apple). In most demographic datasets, different cells in an array are based on populations of different sizes. Because of randomness, values for cells with small population sizes tend to be more variable than values for cells with large population sizes. When we build probabilistic models based on demographic arrays, we need to take these inter-cell differences into account.

Example 8.14. Table 8.2 shows direct estimates of mortality rates for 30–34 year olds in the 22 counties of Wales in 2014. As discussed in Section 4.9, a direct estimate is a count of observed events divided by the population at risk. In

2014, the number of deaths ranged from 0 in Ceredigion and Monmouthshire to 12 in Rhondda Cynon Taf and Cardiff. The counties in Table 8.2 are ordered from smallest to largest, with the smallest (Ceredigion) having 3,130 people aged 30–34 in 2014, and the largest (Cardiff) having 27,080.

In the smaller counties, observed death rates fluctuate erratically, ranging from 0 per 1,000 to 1.64 per 1,00. In the larger counties, the rates are more tightly bunched around the overall mean of 0.6 per 1,000.

TABLE 8.2
Direct estimates of mortality rates per 1,000 for 30–34 year olds in 22 counties of Wales, 2014.

County	Rate	County	Rate
Ceredigion	0.00	Vale of Glamorgan	0.96
Isle of Anglesey	1.64	Flintshire	0.92
Merthyr Tydfil	1.29	Bridgend	0.23
Monmouthshire	0.00	Neath Port Talbot	0.34
Blaenau Gwent	0.72	Wrexham	0.67
Denbighshire	0.45	Newport	0.42
Torfaen	0.56	Carmarthenshire	0.73
Conwy	0.36	Caerphilly	0.61
Powys	1.05	Rhondda Cynon Taf	0.81
Pembrokeshire	0.84	Swansea	0.61
Gwynedd	0.48	Cardiff	0.44

□

When building a probabilistic model involving cells in a demographic array, we need to allow different cells to have different levels of variability, depending on the corresponding population size. Cells based on larger populations, with lower levels of variability, contribute more to the common pool of information than cells based on smaller populations.

Example 8.14 (continued). We estimate super-population death rates for each of the 22 Welsh counties. For a definition of super-population rates, see Section 4.9.

Our preferred approach, depicted in Panel (a) of Figure 8.16, is to treat the counties as exchangeable, but not identical. Each county has its own underlying mortality rate, but that rate is drawn from a common distribution. The estimate for each county not only uses its own data, but also uses data on other counties, so there is some pooling of information. The balance between using its own data and using other units' data depends on population size for that cell. Units with smaller populations borrow more heavily from the remaining units.

We also consider two alternative, contrasting approaches. Under the 'no pooling' approach, depicted in Panel (b), each county is treated as completely distinct and non-exchangeable. The rate for each county is estimated using the county-specific data, separately from other counties. No information is pooled across counties.

Under the second approach, which lies at the opposite end of the spectrum, counties are not just exchangeable but identical. Each county is assumed to share the same underlying mortality rate. This common rate is estimated using data on all counties. There is complete pooling of information across counties.

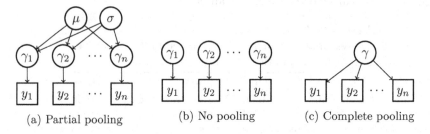

(a) Partial pooling   (b) No pooling   (c) Complete pooling

FIGURE 8.16: Three approaches to estimating mortality in Welsh counties.

In Figures 8.16, the model with partial pooling is

$$y_r \sim \text{Poisson}(\gamma_r w_r)$$
$$\log \gamma_r \sim \text{N}(\mu, \sigma^2)$$
$$\mu \sim \text{N}(0, 10^2)$$
$$\sigma \sim t_7^+(1),$$

the model with no pooling is

$$y_r \sim \text{Poisson}(\gamma_r w_r)$$
$$\log \gamma_r \sim \text{N}(0, 10^2),$$

and the model with complete pooling is

$$y_r \sim \text{Poisson}(\gamma w_r)$$
$$\log \gamma \sim \text{N}(0, 10^2).$$

Figure 8.17 shows estimates produced by the three approaches. The figure includes 95% credible intervals. We discuss credible intervals in Section 9.2, but for the moment it is enough to note that, under the model assumptions, a 95% credible interval for mortality rate has a 95% chance of containing the true mortality rate. The wider a county's credible interval, the more uncertainty there is about the county's mortality rate.

The no-pooling estimate for each county is centered on the observed rate for that county. The complete-pooling estimate is centered on the observed rate for the whole of Wales, regardless of the local observed rate. The partial-pooling estimate is a compromise between the two, with local observed rate exerting less influence on the pooled estimate in small counties than in large counties.

With no pooling, credible intervals are in general much wider (implying much greater uncertainty) in small counties than in large counties. With complete pooling, each county's credible interval is identical. With partial pooling, credible intervals are also in general wider in small counties than in large counties, but the differences are much less dramatic than in the no pooling case. In general, for each county, the credible interval in the partial pooling case is narrower than in the no pooling case, and wider than in the complete pooling case.

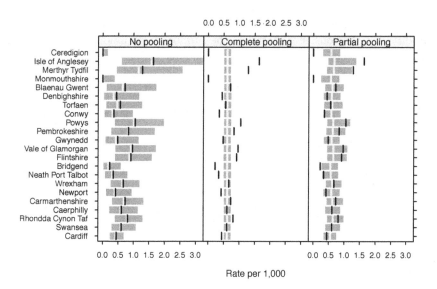

FIGURE 8.17: Estimated super-population mortality rates for 30–34 year olds in Wales counties in 2014, under no pooling, complete pooling, and partial pooling. The counties are ordered from top to bottom by increasing population size. The dark gray bands represent 95% credible intervals, and the light gray lines represent point estimates. The black lines represent the observed mortality rates. □

The tendency for partial-pooling estimates to be pulled towards some central value, such as the overall observed rate across all counties, is referred to as 'shrinkage' or 'regularization'. Shrinkage is common in Bayesian models, though it is found in some non-Bayesian models as well. Mathematical theory and empirical studies both show that shrinkage usually results in estimates that are closer to the true values and more precise. Section 9.7 gives a simple illustration of this principle.

8.7   Hierarchy

The models that we have considered so far have had varying numbers of layers. The random walk model in Figure 8.14 has a single layer. The value for each unit depends only on the value of the unit preceding it. The no-pooling and complete-pooling models in Figure 8.16 each have two layers. The value for each unit in the lower layer depends on a parameter in the layer above. The partial-pooling model has three layers. The value for each unit in the lowest layer depends on a parameter in the layer above, which is governed by parameters in a layer about that. In later chapters, we will encounter models with four, five, or more layers.

The layers in the diagrams of models often correspond to conditional probability distributions. For instance, the random walk model of Figure 8.14, which has a single layer, is composed of terms such as

$$p(y_i|y_{i-1}, \sigma^2);$$

the no-pooling model of Figure 8.16, which has two layers, is composed of terms such as

$$p(y_i|\gamma_i, w_i) \text{ and } p(\gamma_i);$$

and the partial-pooling model of Figure 8.16, which has three layers, is composed of terms such as

$$p(y_i|\gamma_i, n_i), \quad p(\gamma_i|\mu, \sigma^2), \quad \text{and} \quad p(\mu, \sigma^2).$$

Multi-layer Bayesian models are known as Bayesian hierarchical models, or hierarchical Bayes models (abbreviated to HB). In modern Bayesian analysis, virtually all models are hierarchical.

Hierarchy is an essential ingredient of partial pooling models. To obtain partial pooling, we need to treat the quantity we are estimating as a draw from a higher-level distribution. This is why the partial-pooling model in Example 8.14 has three layers, while the no-pooling and complete-pooling models only have two.

More generally, hierarchical models allow statisticians to model multiple sources of variability. This can be important in practice. Traditional demographic models, for instance, allow average levels for demographic rates to vary across geographical regions, but usually do not allow age-profiles to vary. Ignoring variation in age-profiles can be a serious omission. It can, for instance, disguise the fact that some women start their childbearing at different ages in different regions. With hierarchical models, it is relatively easy to capture variation in age profiles, by treating each region's age profile as a draw from a distribution.

When using disaggregated data, hierarchical models can grow large, with thousands, or even millions, of parameters. Large models pose computational challenges. However, understanding these large models is easier than might be thought. They are generally made up of many submodels, each of which is relatively simple. A complicated model for fertility rates, for instance, might contain a submodel describing how births are generated from fertility rates and exposures, a submodel describing how levels vary across regions, a submodel describing how levels vary across age, and so on. To make sense of the overall model, we can go one-by-one through the submodels, and then see how they all fit together.

8.8   Incorporating External Information

When doing demographic modeling, we often have external information about the phenomenon we are modeling, beyond what is included in the data itself. When modeling mortality change, for instance, we may know from our reading of the demographic literature that life expectancies in developed countries generally increase by 1–3 years per decade, and that mortality rates among infants tend to be much higher than those for children aged 1–4.

If we do not build this sort of external information into our model, then our model may perform worse than it would have otherwise. The estimates may be slightly too high or low, for instance, or the credible intervals may be wider than they needed to be.

The loss of accuracy in the estimates depends on the quality of the external information, and on the amount of information contained in the data. If we have a large dataset with information on all the questions of interest, then the dataset along may already be able to provide accurate answers. In this case, the extra effort required to incorporate the external information may not be worthwhile.

Care is required, however, when assessing the amount of information contained in a dataset. The dataset needs to be large in a relevant way. The Chinese 2010 population census, for instance, has over 1.3 billion records. The sample size for any combination of age and sex is so large that a model of, say, educational attainment by age and sex would (if we assume that the census data are accurate) have little need for external information. This situation would be different, however, if we were modeling change in educational attainment over time. The number of time periods sampled by the 2010 census is very small: one. And even if we were to use all the modern Chinese censuses, the number of periods sampled would only increase to six. Incorporating external information into an analysis can be useful, then, even with huge datasets.

Bayesian demographers wishing to incorporate external information into their models have a number of options.

8.8.1   Priors

The most distinctively Bayesian way of incorporating external information is to put it into a prior. Priors encoding strong substantive information about the quantity in question are known as 'informative' priors. If we were forecasting immigration, and we knew that the country had just elected a government with a strong anti-immigration stance, then we might use an informative prior that gave a high weight to short-term reductions in numbers. If we were modeling fertility rates, and we expected age-profiles to be smooth, then we might use a prior that gave a low weight to large differences between neighbouring age groups.

'Uninformative' priors, in contrast, try to say as little as possible about a quantity. They consist of claims such as "this quantity is between negative and positive infinity" or "this quantity is a positive integer". Uninformative priors are convenient, in that they require minimal effort to specify. They are often also presented as being more objective than informative priors. Uninformative priors are, however, rather strange if interpreted literally. An uninformative prior for average births per woman, for instance, may imply the average is just as likely to be between 1,000 and 1,010 as between 1 and 10. Uninformative priors general yield sensible posterior distributions when there are enough data. But when estimating a demographic array with many dimensions, there is no guarantee that the data will be enough.

In response to these problems, many Bayesian statisticians have begun to adopt 'weakly informative' priors. A weakly informative prior incorporates external information, but in a way that understates what we actually know about the likely range of values. For instance, a weakly informative prior for average births per woman might place moderate probability on values between 1 and 10, but low probabilities on values smaller than 0.2 or larger than 25. Weakly informative priors are only slightly less convenient than uninformative priors. Typically, specifying the order of magnitude is enough: saying, for instance, that life expectancy could be anywhere between 20 years and 200 years, with a very low chance of being smaller than 1 year or larger than 1,000 years. In return for the small amount of effort required to specify an order of magnitude, we obtain much better performance when datasets are small or noisy.

We have found informative and weakly informative priors to be extremely useful in our own demographic modeling. In particular, we have found it useful to put informative or weakly informative priors on standard deviation terms in models. A standard rule of thumb in statistics is that estimating standard deviations is much harder than estimating means. At the same time, variability is something that it is possible to make broad statements about. In the absence of wars or epidemics, for instance, demographers would expect mortality rates to vary by only a few percentage points from year to year. With a little arithmetic, statements such as this can be used to set values for informative or weakly informative priors on standard deviations. We use arguments of this type to formulate priors on standard deviations later in the book. Examples include Sections 11.7.3, 16.4.1, and 18.6.2.

8.8.2 Covariates

Suppose that we are building a model of regional migration flows, and we observe that areas with universities attract disproportionately large numbers of young migrants. We could incorporate this observation into our model by setting up an informative prior that favoured high migration rates in areas where we knew there was a university. Alternatively, we could incorporate information on universities into the model via a covariate. We might, for instance, include a covariate that took a value of 1 if an area had a university and 0 otherwise. Or, if we had access to the relevant data, we might use a covariate that measured the proportion of university-age people in each area who were enrolled in full-time study.

Covariates are more transparent and easier to automate than informative priors. They are, in general, a good way to bring in extra information. One sort of problem where they do not generally help, however, is forecasting quantities that vary over time. If we were to find, for instance, that fertility rates rose and fell with GDP growth, we might expect that this relationship would help us to forecast future fertility rates. But to make use of the GDP-fertility relationship, we would need to forecast future GDP growth. This would be at least as difficult as forecasting future fertility rates.

8.8.3 Embedding the Model in a Larger Model

In Example 8.14, we pool information across counties of Wales to improve the estimates of mortality rates for 30–34 year olds in 2014. The strategy of pooling of information can be taken a step further. Many of the same county-to-county differences in health risks that explain geographical variation in mortality rates for Welsh 30–34 year olds are likely to apply to 35–39 year olds, and even to 0–4 year olds or 80–84 year olds. By simultaneously estimating county-level mortality rates across many age groups, we can allow estimates of regional variation for other age groups to inform our estimates of regional variation for 30–34 year olds. Similarly, by adding a time dimension to the model, and pooling information across several years, we could allow estimates of regional variation in earlier years or later years to inform our estimates of regional variation in 2014.

In the Welsh example, we would, in effect, be embedding our model for regional variation among 30–34 year olds in 2014 in a larger model of regional variation for multiple age groups and multiple time periods. The technique of embedding the model of interest into a larger model is very general. The objective, as with the Welsh example, is to use information on similar units to improve estimates for the units of interest. If, for instance, we were trying to estimate and forecast fertility rates in an African country, we might fit a hierarchical model to other African countries. If we were trying to estimate hours worked for one occupation, we might fit a hierarchical model for a range of similar occupations.

Whenever we embed a model within a larger model, we are implicitly or explicitly making assumptions about exchangeability. We are assuming, for instance, that regional differences for Welsh people of different ages are exchangeable, or that African countries are exchangeable. As usual, the assumptions about exchangeability can be made more defensible by dividing units into groups, by adding covariates, or by extending the models in other ways.

8.9   References and Further Reading

The data on the sex ratio at birth in England in Wales were obtained from the *Birth Summary Tables, England and Wales 2014* on the UK Office for National Statistics website on May 15, 2016. The data on deaths and population at risk for Welsh 30–34 year olds were obtained from the datasets *deathsarea2014tcm77431322* and *MYE2_population_by_sex_and_age_for_local_authorities_UK_2014*, also on the UK Office for National Statistics website, on October 21, 2016. The data on emigrants from Iceland were obtained from the database *External migration by sex, age and citizenship, 1986-2015*, on the Statistics Iceland website, on January 21, 2017.

Gelman et al. (2014) have detailed discussions of exchangeability and hierarchical models. Gelman and Hill (2007) emphasize the contrast between no-pooling, partial-pooling, and complete-pooling models. The local-level model, and extensions, are described in Prado and West (2010, ch. 4). O'Hagan et al. (2006) is the standard reference on using expert judgement to construct informative priors. The documentation for the Bayesian modeling language *Stan* has advice on default priors (*github.com/stan-dev/stan/wiki/Prior-Choice-Recommendations*), which we have followed in our own software.

9

Bayesian Inference and Model Checking

This chapter discusses the second and third steps of a Bayesian analysis: inference and model checking. We look first at computing, summarizing, and transforming posterior distributions. We then discuss missing data and forecasting, and review techniques for checking a Bayesian model.

9.1 Computation

By combining the formulas for the individual components that make up our probabilistic model, we can derive a mathematical formula for the posterior distribution. But, in most cases, this formula is difficult to use, in that we cannot easily derive probabilities or summary measures from it. When this happens, Bayesian statisticians use computer simulation to generate a large sample of draws from the posterior distribution.

Example 9.1. The upper panels of Figure 9.1 show the distributions obtained by randomly generating a sample of 10, a sample of 100, and a sample of 1,000 from a normal distribution with mean 0 and standard deviation 1. The larger the sample, the closer the sample-based distribution lies to the true distribution.

The lower panels of Figure 9.1 show samples obtained from a more complicated distribution. The distribution is obtained from taking a draw from a normal distribution in which the mean is 0 and the standard deviation follows a half-t distribution with seven degrees of freedom and scale 1. This distribution is difficult to work with mathematically, but is easy to sample from. We simply draw a value σ from the half-t distribution, and then draw a value from a normal distribution with mean 0 and standard deviation σ.

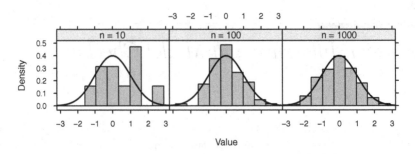

(a) Normal distribution

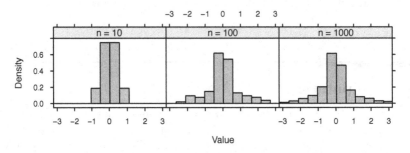

(b) Extension of normal distribution

FIGURE 9.1: Samples from normal distribution and extension of normal distribution. The normal distribution has mean 0 an standard deviation 1. The extension is a mixture of normal distributions with mean 0 and standard deviation drawn from a half-t distribution with seven degrees of freedom and scale 1. The samples have sizes 10, 100, and 1,000.   □

The standard way to obtain a sample from a posterior distribution is to use a set of techniques known as 'Markov chain Monte Carlo' (MCMC). The basic idea of MCMC is to start with a value lying somewhere within the possible range, and then generate a series or 'chain' in which each new value is generated randomly given the preceding value. The new values are generated in such a way that, at each iteration, the chain tends to move towards values that have high posterior probabilities.

In most applications, initial values are chosen through some sort of approximation, and are not a genuine draw from the posterior distribution. When the rules for generating new values are set up properly, the fact that each iteration is random means that a chain eventually forgets its starting point. If we discard early draws in the chain, referred to as burn-in, then the remaining draws will be representative of the posterior distribution.

The amount of time that the chain spends at each value θ is proportional to the posterior probability of θ. If, for instance, there were only 100 possible values for θ, and the 32nd possible value had twice the posterior probability of the 31st, then the chain should visit the 32nd possible value about twice as often as it visits the 31st.

Example 9.2. Figure 9.2 shows two chains generated using MCMC methods. The true posterior distribution is normal with mean 0 and standard deviation 1. The chains appear to have forgotten their starting point by first few iterations. Altogether, 68% of iterations 50–500 from chain 1 and 69% of iterations 50–500 from chain 2 lie within the interval (-1, 1). The theoretical expected proportion is 68%. Running the two processes for longer to boost sample size would reduce random variation and produce an even better approximation of the true distribution.

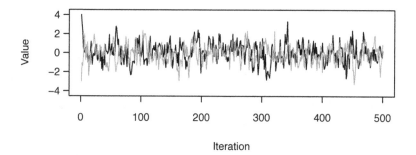

FIGURE 9.2: Two chains generated using MCMC. The first (shown in black) starts at 4, and the second (shown in gray) starts at -3. The posterior distribution is normal with mean 0 and standard deviation 1. □

In Example 9.2 we know the correct answer, so we know whether the MCMC process has produced draws from the right distribution. In realistic problems, we do not know the answer. We therefore need methods for assessing the performance of MCMC without it. The most common technique is to run multiple chains, starting from different initial values, and look for the point at which they all appear to be drawing from the same distribution. We assume that this distribution is the correct one. Chains that appear to be drawing from the same distribution are said to have converged. Bayesian statisticians have developed formal measures to judge convergence of the chains.

All of the models we consider in this book have been implemented in our R packages, which look after most of the computational details. It is not necessary to have a deep understanding of MCMC to use the methods. A little knowledge is, nevertheless, helpful for making the calculations run faster, or for diagnosing problems when something goes wrong. We provide more details, plus examples, on the book's website www.bdef-book.com.

9.2   Summarizing the Posterior Distribution

9.2.1   Summary Measures

Posterior distributions are complicated things. Typically, we do not need all this complexity to answer the question at hand. Instead, we need one or more summary measures.

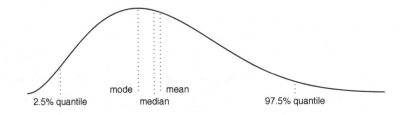

FIGURE 9.3: Quantities often used to summarize a posterior distribution.

End users often want a point estimate, that is, they want a single-number answer such as "the population of China is 1.4 billion", "the census missed 4% of the population", or "the birth rate is 23 per 1,000". Bayesians have various conventions on how to reduce a posterior distribution to a single point estimate. Three of the most popular options—the mode, median, and mean— are shown in Figure 9.3.

The three options measure three different types of centrality. The mode is the value with the highest probability (or probability density). The median is the middle of the distribution, in the sense that half of all values are higher, and half are lower. The mean is the average value of the unknown quantity. Compared with the median, it gives a higher weight to values far from the center of the distribution. The fact that the mean is more sensitive to extreme values can be an advantage or a disadvantage, depending on the application.

For some problems, however, a central value may not be the most useful single-number summary. Consider, for instance, a Ministry of Education planner trying to decide on the appropriate size for a new school. The cost of building a school that is too large and the cost of building a school that is too small may be quite different. Having under-used classrooms is wasteful, but adding extra classrooms to an existing school may be much worse. In such cases, the planner is likely to find an upper bound on the future numbers more useful than a central value.

The 97.5% quantile shown in Figure 9.3 is one possible upper bound. The 97.5% quantile is the point 97.5% of the way through the posterior distribution, starting from the bottom. If Figure 9.3 was a posterior distribution for the future school roll, for instance, and the 97.5% quantile was located at the value 1,900, then we could infer that the future school roll had a 97.5% containing fewer than 1,900 students and a 2.5% chance of containing more.

If the Ministry of Education planner thought that a 2.5% chance of being too small was appropriate, then she might wish to work on the assumption that the school would serve 1,900 students. If she wanted to reduce the chance of being too small, then she could use a higher quantile, such as 99%. If she was willing to tolerate a higher chance of being too small, in return for a reduced chance of being too large, then she might use a lower quantile, such as 90% or 80%. The analysis could be extended to bring in other costs and benefits. But the essential point is that the posterior distribution makes this sort of analysis possible.

Sometimes there are many end users for an analysis, and it is not possible to predict what all their needs will be. In such cases, Bayesians typically provide a general-purpose summary. The standard approach is to give a point estimate, such as a mode, median, or mean, plus one or more 'credible intervals'. A X% credible interval for an unknown quantity is a pair of numbers that encloses X% of the posterior distribution for that quantity. For instance, if a 90% credible interval for the birth rate is 0.034–0.041, then 90% of the posterior distribution lies between the values 0.034 and 0.041. In the distribution in Figure 9.3, the 2.5% and 97.5% quantiles form a 95% credible interval (97.5%− 2.5% = 95%.)

For historical reasons, 95% is the most common value for a credible interval. However, there is nothing magic about the number 95%. Economic forecasters often use 80% intervals. In later chapters we typically show 95% intervals and 50% intervals.

Whatever the values they extract from a posterior distribution, analysts have a professional responsibility to provide measures of uncertainty. If we give a point estimate, for instance, we should accompany it with something like a credible interval.

A point estimate plus a measure of uncertainty conveys more information than a point estimate on its own. Unfortunately, end users do not always embrace the extra information with the enthusiasm that theories of rational decision-making say they should. Sometimes this is because end users like the simplicity of a single value. Sometimes it is because they have an aversion to uncertainty. Sometimes it is because the analyst has not done a good job at explaining the uncertainty measures.

Whatever the cause, it is important that modelers persist in searching for ways to draw users' attention to uncertainty. Moreover, it is important that uncertainty measures be quantitative. Psychologists have shown that people interpret qualitative terms such as "probable" and "unlikely" in dramatically different ways, leading to misunderstandings and bad decisions. Psychologists have also shown that people without statistical training can interpret quantitative measures of uncertainty correctly, provided the measures are well-chosen and clearly explained.

9.2.2   Calculating Posterior Summaries

If we have run our MCMC simulation to convergence, and have accumulated a sample from the posterior distribution, then calculating summary measures is easy. We use the fact that, for a sufficiently large sample,

a summary measure for the posterior distribution	$\approx$	the same summary measure calculated from the posterior sample

(where $\approx$ means "is approximately equal to"). For instance, the mean for the posterior distribution is approximately equal to the mean of the sampled values, the 2.5% quantile for the posterior distribution is approximately equal to the 2.5% quantile for the sampled values, and so on.

Example 9.3. Table 9.1 shows a hypothetical sample from a posterior distribution for mortality rates. We would obtain a sample like this by setting up a model, plugging in the data, and doing MCMC (or having the software do it for us).

TABLE 9.1
Hypothetical sample from a posterior distribution

Draw	Females	Males
1	0.023	0.034
2	0.029	0.032
3	0.025	0.035
$\vdots$	$\vdots$	$\vdots$
1000	0.024	0.037

To obtain the mean of the posterior distribution for female mortality rates, we would calculate the mean of the sampled values, i.e. $(0.023+0.029+0.025+ \cdots + 0.024)/1000$. □

Calculating some summary measures, such as quantiles, can be tricky, but thankfully standard *R* functions such as `quantile` look after the details for us.

The size of the sample required to get good approximations varies from problem to problem. A general rule of thumb is that measures of the center of the distribution, such as means, medians, and modes, require smaller samples than measures of the tails of the distribution, such as 2.5% and 97.5% quantiles. In a well-behaved model, where the MCMC simulations are working efficiently, it may be possible to get good approximations of the mean with a sample of 100 or so, and to get good approximations for quantiles such as 2.5% with a sample of 1000 or so.

9.3 Derived Distributions

9.3.1 Posterior Distribution for Derived Quantities

It is surprisingly easy to extend the procedures in Section 9.2 to deal with new unknown quantities derived from unknown quantities in our posterior distribution. We simply calculate the derived quantities once for each draw in our sample, and then summarize the results.

Example 9.3 (continued). In addition to estimating female and male mortality rates, we would like to estimate a derived quantity: the ratio between the female rate and the male rate. To do this we calculate the ratio for each draw from the posterior distribution: $0.023/0.034 = 0.676$, $0.029/0.032 = 0.906$, and so on, as shown in Table 9.2. We then summarize the values $0.676, 0.906, \ldots, 0.649$ in exactly the same way that we would summarize the values from posterior distribution itself. For instance, to obtain the mean ratio, we would calculate $(0.676 + 0.906 + \cdots + 0.649)/1000$.

TABLE 9.2

Calculating the ratio of female rates to male rates

Draw	Females	Males	*Females/Males*
1	0.023	0.034	*0.676*
2	0.029	0.032	*0.906*
3	0.025	0.035	*0.714*
⋮	⋮	⋮	⋮
1000	0.024	0.037	*0.649*

□

The ability to easily make inferences about derived quantities is extremely useful in practice. In fact, for many applications, it is one of the crucial advantages of Bayesian methods.

The life expectancies for Māori shown in Figure 1.3 in Chapter 1 were calculated in essentially the same way as the ratios in Example 9.3. First we obtained a sample of mortality rates at each age. Next we fed each set of mortality rates into the standard formula for calculating life expectancy. Finally, we obtained quantiles to display in the graph.

One further example of a derived quantity, which is greatly valued by policy analysts and administrators, is rankings.

Example 9.4 (continued from Example 8.14). Health administrators often want to know which regions have the lowest mortality rates and which have the highest, as a way of measuring equity or health system performance. Figure 9.4 shows a ranking of this kind, generated from the posterior distribution for the partial-pooling model in Figure 8.17. Figure 9.4 shows, for each of the 22 counties of Wales, the probability that the county has the lowest mortality rate for 30–34 year olds, and the probability that it has the highest mortality rate.

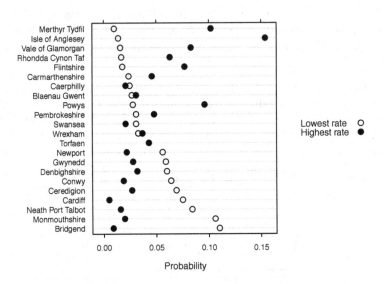

FIGURE 9.4: Probability of having the lowest and highest mortality rates for 30–34 year olds in Welsh counties, 2014.

Deriving the probabilities in Figure 9.4 is straightforward. For each draw from the posterior distribution for mortality rates, we identify which county has the lowest mortality rate and which has the highest. To obtain the probability that Ceredigion County, for instance, has the lowest mortality rate, we calculate the proportion of draws in which Ceredigion ranks lowest. To obtain the probability that Ceredigion County has the highest mortality rate, we calculate the proportion for draws in which Ceredigion ranks highest.

Although Figure 9.4 shows the Isle of Anglesey to have the highest probability of the ranking worst, it also shows that this probability is only about 0.16. Any conclusions about especially high mortality rates on the Isle of Anglesey should be tentative at best. The probabilities indicate that data at hand are not sufficiently informative to permitat us to identify bad performers or good performers with any sort of confidence.

According to Figure 9.4, Monmouthshire has a slightly higher probability than Bridgend of having the worst mortality rates. As can be seen in Table 8.2, however, the direct estimate for Monmouthshire is 0 per 1,000, while the direct estimate for Bridgend is 0.23 per 1,000. What is happening?

Despite appearances, giving Bridgend a lower probability of bad mortality rates makes sense. Bridgen has a larger population than Monmouthshire. The relatively low number of deaths in Bridgen is therefore less likely to be an artefact of random variation, and more likely to reflect the true underlying rates. □

9.3.2 Posterior Predictive Distribution

Later in the book, to do forecasting (Section 9.5) and model checking (Section 9.6.3), we need to work with quantities that depend on unknown quantities in our posterior distribution, but are not determined completely by them. The probability distribution for such quantities is referred to as the posterior predictive distribution. For instance, a posterior predictive distribution could describe the death counts that we would observe if we could somehow replay history, while keeping the chance of having a death the same.

Example 9.4. In Table 9.3 we use the posterior distribution from Table 9.1 to derive a posterior predictive distribution for female death counts in a population that contained an average of 100 women over a one-year period. We derive the first value by drawing from a Poisson distribution with rate 0.023 and exposure 100; we derive the second value by drawing from a Poisson distribution with rate 0.029 and exposure 100; and so on. The sampled counts are shown in Table 9.3. We can summarize the counts 3, 2, $\cdots$, 4 in exactly the same way that we summarize the values from posterior distribution itself. □

TABLE 9.3

Posterior predictive distribution for death counts for females

Draw	Death rate	Exposure	Death count
1	0.023	100	*3*
2	0.029	100	*2*
3	0.025	100	*3*
⋮	⋮	⋮	⋮
1000	0.024	100	*4*

The posterior predictive distribution combines two sorts of uncertainty: (i) uncertainty about the true value of unknown quantities in our posterior distribution, and (ii) uncertainty in how the quantities of interest are generated given the value of unknown quantities in our posterior distribution.

The posterior predictive distribution is typically used for finite-population quantities that could potentially be observed. As shown in Section 12.11, however, it can also be used for super-population quantities.

9.4   Missing Data

Real datasets often have missing values. Respondents skip questions, published tabulations are incomplete, and administrative systems leave out parts of the target population. Probabilistic models in demography, like probabilistic models in other fields, need to be able to deal with gaps in data.

The Bayesian approach to missing data is to treat the missing values as unknown quantities, and include them in the joint probabilistic model and the posterior distribution along with other unknown quantities. The estimates for the missing data depend on the specifics of the model, and particularly assumptions about exchangeability.

Example 9.5 (continued from Example 8.14). We examine what would happen if our data on deaths among Welsh 30–34 year olds were missing values for the counties of Monmouthshire and Newport. Under the no-pooling model, each county is regarded as unique, and information cannot be shared across counties. Without the ability to share information, we cannot make any progress on the missing values. We therefore have to give up on the no-pooling model.

The complete-pooling model, in contrast, deals easily with missing data. Under the complete-pooling model, every county has the same mortality rate. This rate can be estimated from the 20 counties with data, and then applied to the two counties without data. The left-hand panel in Figure 9.5 shows the result.

Under the partial-pooling model, each county has its own rate, but this rate is drawn from a distribution that is shared with other counties. The parameters of the shared distribution can be estimated from the 20 counties with data, and then used to derive rates for the two counties without data. The right-hand panel of Figure 9.5 shows the results.

Under both models the rates for Monmouthshire and Newport are centered on the overall average. With the complete-pooling model, the rates for these two counties are no more uncertain than those for counties with data. With the partial-pooling model, however, the two counties with missing data have slightly wider credible intervals than counties with similar population sizes.

By combining the estimated rates with the exposure measures, we can derive posterior distributions for the missing values, as shown in Figure 9.6. The distributions obtained from the partial-pooling model are slightly more spread out than those from the complete-pooling model, implying greater uncertainty.

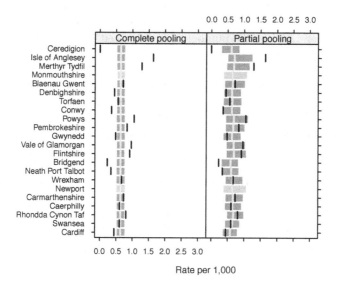

FIGURE 9.5: Estimated super-population mortality rates for 30–34 year olds in Welsh counties in 2014, with missing data for Monmouthshire and Newport counties. Estimates are shown for complete-pooling and partial-pooling models. The gray bands represent 95% credible intervals, and the light gray lines represent posterior medians. The black lines represent the observed death rates.

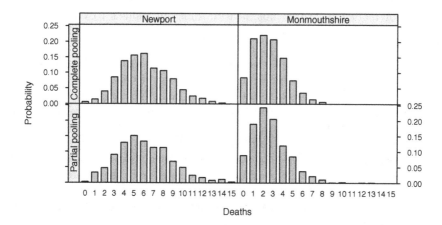

FIGURE 9.6: Imputed death counts for Monmouthshire and Newport, under the complete-pooling and partial-pooling models. □

One traditional approach to missing values is to fill them in with some sort of best guess, and then treat the imputed values as known, just as if they were directly observed. With Welsh counties, for instance, if we apply a mortality rate of 0.5 per 1,000 to the number of people aged 30–34 in Monmouthshire, the expected number of deaths is approximately two. For Newport, the expected number is five. It is tempting to simply assume that Monmouthshire had two deaths and Newport five, and then proceed as normal. However, if we consider Figure 9.6 to represent the true uncertainty about the number of deaths, then fixing the deaths at two and five would mean we were overstating our actual knowledge, leading to estimates that were spuriously precise.

The basic complete-pooling and partial-pooling models both assume complete exchangeability between units. This may be appropriate for Welsh counties, but is not necessarily true in general. In fact, we may often suspect that units that have missing values differ systematically from units that do not. For instance, if regions that fail to supply data to the national statistical office tend to be poorly administered, and if poorly administered regions tend to have higher mortality, then regions with missing data will tend to have higher mortality.

As with exchangeability assumptions in general, assumptions about the exchangeability of units with and without missing data can be made more plausible by dividing the units into groups or by adding covariates. We might, for instance, assume that low-income regions that have missing data are exchangeable with low-income regions that have complete data, while high-income regions that have missing data are exchangeable with high-income incomes that have complete data. The assumption that units with missing values are exchangeable with units with complete data, after dividing the units into groups or adding covariates, is known as a "missing at random" assumption.

The ability to deal appropriately with missing data opens up new possibilities for modeling. We do not have to limit our models to variables with complete observations. Indeed, we may even include in our models variables for which there are *no* observations.

One demographic example including a variable with no observations is Lexis triangles. As discussed in Section 4.7, to convert between age-based and cohort-based classifications of events, we need to assign each event to a lower or upper Lexis triangle. In the absence of actual data on Lexis triangles, the traditional demographic approach is to assume that exactly half of all events belong to lower triangles, and half belong to upper triangles. This assumption is equivalent to replacing missing values with best guesses and treating them as data. As we have seen, treating best guesses as if they were data can lead to estimates that are spuriously precise. A more satisfactory approach is to treat the Lexis triangles as missing data, and to infer them as part of the estimation process, as we do in Chapter 19. We do not escape the need to make assumptions, such as an assumption that rates in lower triangles were similar to those in upper triangles. But at least the assumptions are transparent, and we account properly for uncertainty. An example of this is given in Section 19.2.3.

9.5 Forecasting

Far better an approximate answer to the right question, which is often vague, than an exact answer to the wrong question, which can always be made precise.

— John Tukey

Demographers often treat forecasting and estimation as distinct activities. From a Bayesian perspective, however, forecasting is just a type of estimation with missing data.

Example 9.6. Consider the position of two Bayesian demographers modeling emigration from Iceland, summarized in Figure 9.7. The first demographer is conducting her analysis at the end of 2010. She has data for the period 1986-2010, though with missing values for 1994–1998, and would like to estimate the underlying emigration rates for the whole period. The second demographer is conducting her analysis at the end of 2005. She has complete data for the period 1986–2005, and would like to forecast rates for 2006–2010, based on estimates for 1986-2005. Both demographers use local-level models.

The data available to the two demographers are shown in the top panels of Figure 9.7, and the results of their analyses are shown in the bottom two panels.

Both demographers in Example 9.6 must obtain rates for years in which there are no data. The reasons for the data being missing are different: in forecasting case, the data are missing because the events in question have not occurred yet. But both sets of inferences are essentially missing data problems.

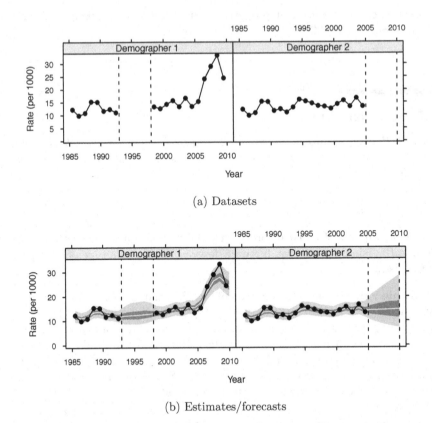

(a) Datasets

(b) Estimates/forecasts

FIGURE 9.7: Emigration rates for Iceland, 1986-2010. The black lines show total emigrations divided by total population. Demographer 1 is estimating rates for the period 1986-2010, with missing values for 1994-1998. Demographer 2 is forecasting rates for the period 2006-2010, based on estimates for 1986-2005. □

Technically, there are two equivalent ways of doing forecasting with Bayesian models. One way is to include the forecasted quantities as part of the unknown quantities in the joint probabilistic model, and generate draws for them as part of draws from the posterior distribution. The other way is not to include the forecasted quantities in the posterior distribution, generate draws from the posterior distribution first, and then generate draws for the forecasted quantities from their posterior predictive distribution as defined in Section 9.3.2. Both ways of forecasting are fine for short-term forecasting. For long-term forecasting, the first way of forecasting could encounter computational issues, in that the chains starting from different initial values can be slow to convergence.

With forecasting, as with other missing data problems, assumptions about exchangeability play a central role. To forecast with a local-level model, for instance, we must assume that increments in future years will be drawn from the same distribution as increments in past years. In other words, the series being forecast must, on average, take steps of the same size upwards or downwards in future years as it has in past years.

Exchangeability assumptions become increasingly suspect, the further out the forecast extends. Technologies, government policies, social norms, and other influences on demographic dynamics can be expected to look much the same in 5 years as they do now. But these influences, and hence the demographic dynamics, may be quite different in 50 years' time.

As we discuss in subsequent chapters, modelers do have ways of making their forecasts more reliable. They can base their forecasts on long historical time series. They can bring in external information, by embedding the forecast in a bigger model that draws on the experience of multiple populations, or by using informative priors. They can examine the sensitivity of their conclusions to alternative specifications.

Of course, even the most carefully constructed forecasts can be wrong. The demographer in Example 9.6 making an emigration forecast in 2005 could not have foreseen the 2008 financial crisis and ensuing spike in emigration from Iceland. Demographers like to point out that, in the 1930s and early 1940s, virtually no one predicted the European and American baby boom.

Indeed, some people argue that demographers should not produce forecasts at all. Instead, according to this view, demographers should produce projections, in the sense of hypothetical scenarios that merely illustrate what *could* happen. Most population projections contain a disclaimer somewhere in the fine print saying that the projection is only as accurate as its assumptions about future rates, and that it is the users' responsibility to evaluate these assumptions. Similarly, some national statistical agencies have a policy of producing estimates, but not forecasts. These agencies consider forecasts to be too speculative to warrant the title of official statistics.

Some of this skepticism is healthy. It is possible to be overly impressed by the sophisticated mathematics and beautiful graphical displays of modern forecasting methods, and to forget that they are just sophisticated forms of extrapolation. In our view, however, abstaining from forecasting is not the most helpful response. If people planning schools, supermarkets, pension funds, retirement policies, roads or other population-dependent assets or policies do not get the answers they need from statisticians and demographers, they will seek them elsewhere.

Population projections are, in general, exact answers to the wrong questions. They answer questions of the form "what would happen if X?" But what the public actually wants to know is "what is likely to happen?" Answering that question requires a forecast, with appropriate measures of uncertainty. Forecasting is difficult, and all forecasts, including their measures of uncertainty, are necessarily approximate. But at least forecasts answer the questions that were asked.

9.6   Model Checking

9.6.1   Responsible Modelers Check and Revise their Models

We cannot guarantee that our model had captured all the important features of the system under study. The world is infinitely complicated, and our models are not. Even the biggest models require educated guesses about what should be included and what should be left out. These educated guesses often turn out to be wrong.

Sometimes a model's shortcomings are obvious, as when a forecasting model predicts a life expectancy of 5,000 years. But often the problems are more subtle, and come to light only after careful investigation. Good modelers cultivate a skeptical attitude towards their models. They do not stop as soon as they have a model that they can fit and that gives a plausible answer. Instead, they force themselves to search for hidden deficiencies indicating that the model is not serving the purposes for which it was built. If such deficiencies become apparent, then the modelers revise their models. Where necessary, they will go through the check-revise cycle several times.

It is difficult to formulate rules for model checking, since the checks need to be guided by the specifics of the data and the application. However, to give a sense of what is possible, we present two techniques for model checking here, and provide further examples in the remainder of the book.

9.6.2   Heldback Data

If our model has captured the main features of the phenomenon of interest, then it should to a good job of imputing missing data. A common model-checking technique is to deliberately hold back data from a model, and see if the model can indeed perform the imputations.

Example 9.7. We test the ability of a local level model of British sex ratios at birth to predict heldback data. We fit the model to data for the period 1938–2004, and see if it can predict the values for 2005–2014. As can be seen in Figure 9.8, the model performs reasonably well, in that most of the heldback values fall within the 50% credible interval.

□

The more data we hold back from the model, the stronger the test is. However, we cannot hold back too much data, or the model will not have enough observations to form reliable estimates. Heldback data techniques can work well with hierarchical models in which values are cross-classified along several dimensions. We can then see whether the model does a better job of imputing data for some age groups rather than others, for instance, or does a better job in some regions rather than others.

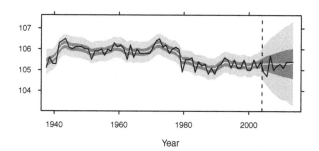

FIGURE 9.8: Using heldback data to check a local level model for UK sex ratios. The solid line represents data on the sex ratio. Data to the left of the dashed line were supplied to the model, and data to the right were held back. The light gray and dark gray bands represent 95% and 50% credible intervals, and the white line represents the posterior median.

Much of the scrutinizing of models that modelers routinely do can be interpreted as a type of informal heldback data check. Say, for instance, that we fit a model of mortality rates, and find that life expectancies for males are higher than life expectancies for females. In virtually all modern populations, the relationship is the other way round: female life expectancies exceed male life expectancies. If we did not explicitly build that fact into our model, then we can be seen as performing a type of heldback data check in which we see if the model will reproduce the expected female-male relationship. The failure of our model to reproduce the expected relationship raises questions about its validity in essentially the same way that the failure to reproduce some heldback death counts would.

We return to the subject of heldback data in Section 12.7.

9.6.3 Replicate Data

Another way to assess a model's ability to capture relevant features of the data is to use the model to generate replicate data that could have been observed instead of the actual data. Each replicate dataset can be drawn from a posterior predictive distribution, as defined in Section 9.3.2. The logic is set out in Figure 9.9. We use the real data to fit the model, and then use the model to randomly generate replicate datasets. If the model is an adequate description of the underlying demographic processes, then the replicate datasets should look like the actual dataset. The replicate and actual data will not look exactly the same, since both are outcomes of random processes. However, the actual data should not diverge systematically from the replicate data, or otherwise appear out of place.

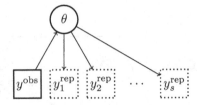

FIGURE 9.9: Model checking using replicate data. The data y^{obs} are used to fit the model θ. The model is then used to generate replicate datasets $y_1^{rep}, y_2^{rep}, \cdots, y_s^{rep}$. If the model is working, the real data should look as if it was drawn from the same distribution as the replicate data.

Humans are adept at spotting subtle differences in visual patterns, so the best way to tell whether the actual data diverge from the replicate data is usually through graphs. It can, nevertheless, be helpful to summarize any observed differences through quantitative measures. Indeed, the process of calculating summary measures can be formalized into procedures that are reminiscent of classical hypothesis testing.

Example 9.7 (continued). Figure 9.10 shows the actual data for UK sex ratios, plus 19 replicate datasets generated from a local level model. The actual data do not look distinctive, except that they possibly have a more step-like appearance than the replicate data. The actual data seems to cluster around one value between 1942 to 1979, then fall abruptly, before clustering around a new level between 1980 and 2014. The sex ratio drops by 1.1 between 1979 and 1980, which is the biggest year-on-year change in any of the 20 datasets in Figure 9.10.

We might consider revising the model so that year-on-year changes were generally small, but were interspersed by occasional large jumps. Whether the extra effort and complexity were worthwhile would depend on the uses to which the model was going to be put.

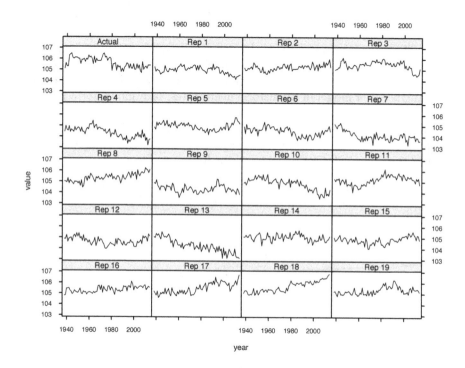

FIGURE 9.10: Using replicate data to check a local level model for UK sex ratios.

□

9.7   Simulation and Calibration*

We use model checking to assess how a particular model performs with a particular dataset. Sometimes, however, we would like to know how a model would perform over a range of different datasets. We would like to identify the sorts of datasets, and hence the sorts of applications, where a model can be expected to give sensible results.

Knowledge about which sorts of models are appropriate for which sorts of problems tends to accumulate naturally within a discipline, as practitioners gain experience with different types of models, and keep track of the models' ability to give plausible answers, survive checks, and provide accurate forecasts. The accumulation of knowledge can be slow and uncertain, however, if there is no gold standard to compare model results against. Even with imaginative and ruthless model checking, for instance, it is difficult to assess the performance of a model of prehistoric death rates, or a model for a rare and difficult-to-diagnose medical condition.

In the absence of actual gold standards to assess models against, an excellent alternative is to generate artificial gold standards, and assess models against them instead. The procedure is as follows:

1. Construct some plausible values for the unknown quantity of interest. For instance, construct some death rates. These represent the 'simulation truth'.

2. Use the constructed values to randomly generate some data. For instance, use the death rates to randomly generate some death counts.

3. Fit the model to the randomly-generated data.

4. Compare the estimates from the model with the simulation truth. For instance, compare the estimated death rates with the 'simulation truth' death rates.

If the model estimates come close to the simulation truth, then confidence in the model increases; if the estimates are far from the simulation truth, then confidence decreases. When only a single comparison is made, however, there are always questions about the role of chance. The particular simulated dataset might have been unusually well suited to the model in question, or unusually poorly suited. To minimize the role of chance, steps 1 to 4 can be carried out multiple times, with a different simulated dataset each time. The model is then judged by its performance across all the datasets.

Repeating steps 1 to 4 many times also allows the statistician to test whether a statistical model is well calibrated. A model is well calibrated if its measures of uncertainty are accurate: if, for instance, a 50% credible interval contains the true value at least 50% of the time, a 95% credible interval contains the true value at least 95% of the time, and so on, while keeping the credible intervals as narrow as possible.

Simulations are particularly valuable for fields, such as Bayesian demography, that are relatively new, and that therefore have little accumulated knowledge on which models work best for which problems. Simulations offer the possibility of speeding up the evaluation process.

Moreover, simulations can be applied to models from any statistical tradition, Bayesian, frequentist, or otherwise. All statisticians agree that models should be accurate and well-calibrated. Models from different traditions can be evaluated against each other using commonly-agreed measures of accuracy and calibration.

Example 9.8. We will use simulation to compare the performance of the no-pooling, complete-pooling, and partial-pooling models from Section 8.6. We use artificial data on births for 10 regions, each of which contains an average of 100 women of childbearing age over a one-year period.

Suppose that the simulation truth follows a partial-pooling model. We generate 1,000 artificial datasets by repeating the following steps 1,000 times:

1. Draw μ^{True} from a $N(0, 1)$ distribution.

2. Generate a set of 10 birth rates $\gamma_1^{\text{True}}, \gamma_2^{\text{True}}, \cdots, \gamma_{10}^{\text{True}}$, by drawing 10 values from a normal distribution with $\mu = \mu^{\text{True}}$ and $\sigma = 0.1$, and then exponentiating them. (Note that the birth rates, after a log transformation, follow a normal distribution.)

3. Generate 10 sets of birth counts, $y_1, y_2, \cdots, y_{10}$, by drawing from Poisson distributions with means $100\gamma_1^{\text{True}}, 100\gamma_2^{\text{True}}, \cdots, 100\gamma_{10}^{\text{True}}$.

We fit three models to each artificial dataset: a no-pooling model, a complete-pooling model, and a partial-pooling model. Specifications of these models are given in Section 8.6. Having fitted the models, we calculate three performance measures. The first is the mean squared error (MSE),

$$\text{MSE} = \sum_{r=1}^{10} (\hat{\gamma}_r - \gamma_r^{\text{True}})^2 / 10, \qquad (9.1)$$

where $\hat{\gamma}_r$ is the posterior median for region r. MSE is a standard way of assessing accuracy. The second measure is the proportion of true γ_r that fall within the corresponding 50% credible intervals. The third performance measure is the width of the 50% credible intervals. If we have two models that are accurate and equally well calibrated, we prefer the model that gives narrower credible intervals.

Results from the experiment are shown in Table 9.4. The no-pooling model and the partial-pooling model have correct coverage levels, but the partial-pooling model has lower MSE than the no-pooling model and has narrower credible intervals. The complete-pooling model is less accurate than the partial-pooling model, and has much poorer coverage, which implies that the narrow credible intervals from the complete-pooling model are unrealistically narrow. The partial-pooling model wins.

TABLE 9.4
Results from a simulation study of the no-pooling, complete-pooling, and partial-pooling models.

| | | Pooling | |
Measure	None	Complete	Partial
Mean squared error × 1000	16	49	13
Coverage of 50% credible interval (%)	49	18	49
Width of 50% credible interval × 1000	15	5	12

□

9.8 References and Further Reading

The John Tukey quote at the start of Section 9.5 comes from Tukey (1962).

Fischhoff (2012) reviews psychological studies showing that lay people benefit from using quantitative measures of uncertainty. Little and Rubin (2002) is a classic book on missing data. Braaksma and Zeelenberg (2015, p199) argue that national statistical agencies should not do forecasts. Bijak et al. (2015) argue that national statistical agencies should do probabilistic population forecasts. Bijak (2010) is a detailed discussion of Bayesian demographic forecasting, focusing on migration in Europe, but with wider implications. Bryant and Zhang (2016) present a model for forecasting migration rates, and subject it to a variety of checks. Rubin (1984) is a classic article on simulation and calibration. Little (2012) discusses the role of simulation and calibration in official statistics.

Part III

Inferring Demographic Arrays from Reliable Data

10

Inferring Demographic Arrays from Reliable Data

In Part III, we work with a single demographic array containing counts such as births or deaths, or means such as health expenditure. We assume that the array has no measurement errors, though it may have some missing values. With most models, we also assume that we have an array measuring exposures or population sizes. The exposures array has no measurement errors and no missing values.

We begin Part III, in this chapter, with an overview of our approach. We fill in the details in subsequent chapters, through three case studies.

10.1 Summary of the Framework of Part III

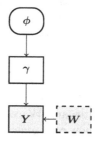

FIGURE 10.1: Inferring an array of rates, probabilities, or means from a single reliable dataset. γ is the array of rates, probabilities, or means; Y is the dataset, also in the form of an array; W is an optional array of exposures; and ϕ holds parameters from the prior model for γ. Datasets Y and W are observed, and γ and ϕ are not.

Figure 10.1, a more mathematical-looking version of Figure 1.4 from Chapter 1, summarizes the framework of Part III. The components are:

Demographic array Y. Data on the demographic series. The data are treated as being free of measurement errors, though they may have missing values. The array almost always contains a time dimension, and also includes other dimensions, such as age, sex, region, income level, or marital status. The elements of Y are finite-population quantities.

Exposure array W. Data on exposures for Poisson models, or numbers of trials for binomial models, with the same dimensions as Y. The data are assumed to have no measurement errors and no missing values. Not all models include W.

Array γ of rates, probabilities or means. An array of parameters with the same dimensions as Y. These parameters are super-population quantities, and appear in the likelihood for Y.

Vector ϕ of parameters from the prior model. The models of Part III have complicated prior models. We use ϕ to denote all the parameters from the prior model.

In Part III, we treat Y as completely free of measurement errors. This is an approximation. It is reasonable if measurement errors in a dataset are small enough that they can be ignored, freeing up time to spend on other issues, such as modeling random variation in event counts, or modeling how rates vary across the dimensions of interest.

If we only want to learn about finite-population quantities (as defined in Section 4.9) and have reliable data with no missing values, then we can dispense with the methods described in Part III. To calculate finite-population death rates, for instance, we simply divide observed death counts by observed population at risk. However, if we do have missing data, then we need models like those of Part III to impute the missing values. If we want to forecast finite-population quantities, we also need a model. Finally, if we want to learn about super-population quantities, then we again cannot escape the use of models.

We treat the elements of Y as random draws from probability distributions. The probability distributions are governed by the elements of γ. For instance, if Y is an array of death counts, then we might use a Poisson model with γ containing the death rates and W containing the exposures. We also have a prior model describing how γ is generated.

The models of Part III generate estimates of the super-population rates, probabilities, or means in γ, as well as parameters ϕ from the prior model. Estimates of γ are usually the main focus. However, estimates of ϕ can also provide insights into the relationship between the outcomes and the various cross-classifying dimensions. If the data Y contains missing values, then these can be estimated too. The estimates take the form of multiple draws from the posterior distribution, as discussed in Section 9.1.

We use a variety of probability distributions to model Y. Table 10.1 gives some examples. As discussed in Section 8.3.1, the Poisson distribution comes in two forms: a rates-exposure form that includes an exposure term, and a counts form that does not. We typically use the rates-exposure form in models of

TABLE 10.1
Examples of components in the framework of Part III

Distribution	Y	W	γ
Poisson (rates-exposure)	Count of births	Person-years lived by women	Fertility rate
Poisson (rates-exposure)	Count of deaths	Person-years lived	Mortality rate
Poisson (counts)	Count of people registered to vote	(None)	Expected number of people registered to vote
Binomial	Count of people with diabetes	Count of people	Probability of having diabetes
Binomial	Count of infant deaths	Count of births	Probability of infant death
Normal	Mean waist circumference	(None)	Expected mean waist circumference

events such as births and deaths, and the counts form in models of population size.

Example 10.1. Figure 10.2 illustrates a hypothetical example in which underlying death rates need to be estimated from death counts. The array of death counts Y, and therefore the array of exposures W and array of death rates γ, has only two dimensions and four cells. Each cell in Y is related to the corresponding cells in W and γ via the equation

$$y_{sr} \sim \text{Poisson}(\gamma_{sr} w_{sr}),$$

where subscript s indexes sex and subscript r indexes region. The prior for the γ_{sr} is simply

$$\gamma_{sr} \sim \text{N}(\mu, \sigma^2).$$

The models we examine in later chapters all have more complicated prior models than this.

The mean parameter μ has a normal prior,

$$\mu \sim \text{N}(0, 1),$$

and the standard deviation parameter σ has a half-t prior,

$$\sigma \sim t_7^+(1).$$

In this example, neither of these priors has any parameters to be estimated.

		region	
		A	**B**
sex	Female	?	?
	Male	?	?

Death rates γ

μ: ?
σ: ?

Parameters ϕ

		region	
		A	**B**
sex	Female	5	8
	Male	7	9

Death counts Y

		region	
		A	**B**
sex	Female	18.3	15.2
	Male	33.8	19.1

Exposure W

FIGURE 10.2: A simple example of a model using the framework of Part III. □

10.2   Applications

The assumption that the data are free of measurement errors is not unusual. Many methods in applied demography, and in applied statistics more generally, make the assumption, though without drawing attention to it.

The models of Part III are useful for learning about demographic rates and propensities when sample sizes are too small for direct estimates to be reliable. The estimation of Māori mortality rates in Section 1.1 is a typical example. There is a subfield within applied statistics devoted to the problem of estimating disaggregated quantities when sample sizes are small, known as small area estimation. The reference to "area" is somewhat unfortunate, as small area estimation models are now applied to problems that are completely non-geographical. The models of Part III can be seen as a special class of small area estimation models, customized for cross-classified demographic data.

Even when sample sizes are large, however, there are settings in which we would still want to use the methods of Part III. By examining the smoothed rates in γ, or the more abstract parameters in the prior model, we may gain insights into the demographic series that we would not gain by looking at direct estimates. The methods of Part III are also needed for dealing with missing data and forecasting.

10.3   References and Further Reading

Gelman and Hill (2007) is an accessible introduction to Bayesian hierarchical models. Rao and Molina (2015) is a book-length review of small area estimation, covering Bayesian and non-Bayesian methods, and Pfeffermann (2013) is an article-length review.

11

Infant Mortality in Sweden

In this chapter we tackle our first application, the estimation and forecasting of infant mortality rates for counties of Sweden. We step through each part of the modeling process: examining the data, formulating a model, scrutinizing model output, testing the model, and forecasting.

In some ways, the application in this chapter is relatively simple. We treat the input data as error-free, and we model variation across only two dimensions: county and time. The application, is nevertheless, substantively important. Rates disaggregated by geographical region can provide evidence on the effectiveness of different regions' health administrations, for instance, or on equity across regions. Moreover, the task of forecasting regional mortality rates is definitely not simple. As we will see, some of the methodological challenges are only partly solved.

11.1 Infant Mortality Rate

The *Encyclopedia of Population* describes the infant mortality rate as "the probability of death in the first year of life", and notes that it is normally calculated by dividing the number of deaths of infants (i.e. children aged less than one year) during a year divided by births during that year. The fact that the infant mortality rate is actually a probability rather than a rate is an endless source of confusion for non-specialists. But even without the terminological difficulties, the practice of dividing deaths in a year by births in that year is somewhat odd, because, as can be seen in Figure 11.1, the population born during a year only partly lines up with the population at risk of dying during that year.

Statisticians and demographers tolerate this misalignment for the sake of simplicity. Unless deaths or births are changing rapidly, it should have little effect on estimated mortality rates. The misalignment does, nevertheless, illustrate the point that likelihoods are approximations, including the most conventional and widely-used ones.

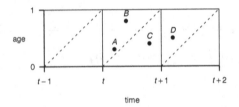

FIGURE 11.1: The population at risk of dying during the first year of life. Dots represent deaths, and dashed lines mark out cohorts. Conventionally, the infant mortality rate is calculated by dividing the number of infant deaths during the period $(t, t+1)$ by the number of births during that period. However, deaths A and B, although occurring during the period $(t, t+1)$, are experienced by the cohort born during the period $(t-1, t)$. Moreover, death D, although occurring during period $(t+1, t+2)$, is experienced by the cohort born during the period $(t, t+1)$.

11.2   Infant Mortality Rates in Swedish Counties

Swedish demographic data are famously accurate. That is not to say that the data are perfect: even Swedes make mistakes when filling out forms. But the data are accurate enough that we can ignore the remaining imperfections, and concentrate on other sources of uncertainty, such as random variation.

Figure 11.2 shows the data we will use to calculate infant mortality rates for 1996–2015, for eight selected counties. (We show only eight to save space.) The first panel shows counts of infant deaths, and the second panel shows counts of births. The third panel shows death counts divided by birth counts. As discussed in Section 4.9, death counts divided by birth counts can be interpreted as the finite-population infant mortality rate, or as a direct estimate of the underlying infant mortality rate.

The counties are ordered in Figure 11.2 by the total number of deaths. The county with the fewest deaths is at the top left, and the county with the most is at the bottom right. In the first two subfigures, each county uses a different vertical scale. (The umlauts and circles in the county names have been omitted. We apologize to any Swedish readers.)

The number of infant deaths is very low in the smallest counties. In Gotland in 2015, for instance, there were no deaths at all. The counts are small enough that random variation dominates. Indeed, random variation is visible even in the data for the most populous county, Stockholm.

Births, in contrast, number in the thousands. The series are correspondingly much smoother. In most cases, there is a slight trend downwards.

The series showing death counts divided by birth counts inherits the variability of the series for death counts, including the tendency for variability to

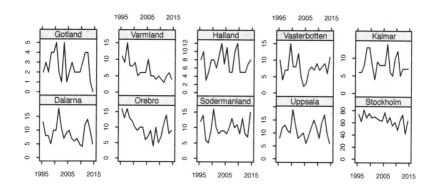

(a) Infant death counts

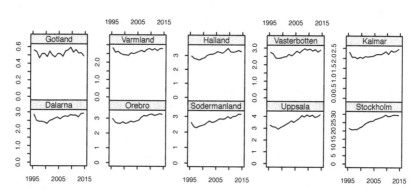

(b) Birth counts (thousands)

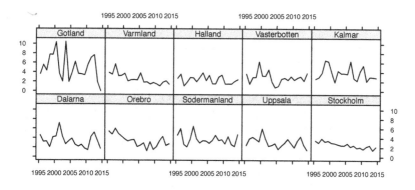

(c) Infant mortality rate (per thousand)

FIGURE 11.2: Infant mortality data for Sweden, for eight selected counties, 1996–2015.

be less in counties with large populations. The ratio of death counts to birth counts appears to be roughly the same across counties, though the noise in the rates makes it difficult to be sure. There is a clear downward trend over time in larger counties, but trends are difficult to discern in the smaller counties.

11.3   Model

11.3.1   Likelihood

We treat the number of deaths occurring in a given county c during a given year t as a draw from a binomial distribution. We use y_{ct} to denote the number of deaths, w_{ct} to denote the number of births, and γ_{ct} to denote the probability that a baby will die during its first year. The model for observed birth counts— that is, the likelihood—is then

$$y_{ct} \sim \text{binomial}(w_{ct}, \gamma_{ct}). \tag{11.1}$$

Our objective is to estimate the underlying infant mortality rates γ_{ct} for the $21 \times 20 = 420$ county-year combinations for which we have data. These 420 probabilities are super-population quantities.

11.3.2   Model for Underlying Infant Mortality Rates

Having described how the number of deaths are expected to vary with the number of births and with the underlying infant mortality rates γ_{ct}, we specify a model for the γ_{ct}:

$$\text{logit}(\gamma_{ct}) \sim \text{N}(\beta^0 + \beta_c^{\text{coun}} + \beta_t^{\text{time}}, \sigma^2). \tag{11.2}$$

Our specification is indirect, in that the left-hand side is a function of γ_{ct}, rather than γ_{ct} itself. The function is the logit function. It allows us to map every point on the probability scale, which is restricted to values between 0 and 1, to the logit scale, where there are no upper or lower bounds. We specify our model on the logit scale, which is much more convenient mathematically than working on the probability scale. The inverse logit function, logit^{-1}, allows us to go from the logit scale back to the probability scale.

The logit function is defined as

$$\text{logit}(p) = \log\left(\frac{p}{1-p}\right),$$

and the inverse logit function is defined as

$$\text{logit}^{-1}(x) = \frac{\exp(x)}{1+\exp(x)}.$$

Figure 11.3 illustrates the effects of a logit transformation. The left-hand graph below shows direct estimates of Swedish infant mortality rates on the original probability scale. The right-hand graph shows the values transformed to the logit scale. The transformed values are all well below 0. This is because the original values were all well below 0.5, which corresponds to 0 on the logit scale.

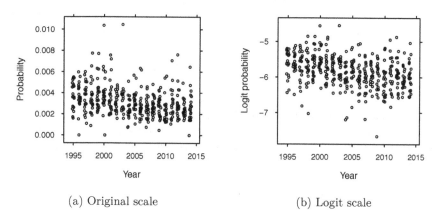

(a) Original scale   (b) Logit scale

FIGURE 11.3: Direct estimates of Swedish infant mortality rates, on original and logit scales.

The β_c^{coun} in Equation (11.2) is a 'region effect'. By including a region effect, we are saying that we expect $\text{logit}(\gamma_{ct})$, and hence γ_{ct}, to vary systematically across counties. In other words, we expect some counties to have consistently high mortality rates and others to have consistently low rates. Similarly, by including a 'time effect' β_t^{time}, we are saying that we expect the average rate across all counties to change from year to year. The β^0 in Equation (11.2) is referred to as the 'intercept'. It controls the average level of the rates.

If we add together the intercept, the region effect, and the time effect for a given county and year, then we obtain a 'predicted' or 'expected' infant mortality rate for that county and year. Let

$$\eta_{ct} = \beta^0 + \beta_c^{\text{coun}} + \beta_t^{\text{time}}.$$

Then η_{ct} is the predicted rate for county c during year t. The value η_{ct} is, however, on the logit scale. To get back to the probability scale, we need to apply the logit^{-1} function. Let

$$\tilde{\gamma}_{ct} = \text{logit}^{-1}(\eta_{ct}).$$

Then $\tilde{\gamma}_{ct}$ is the predicted infant mortality rate, measured on the probability scale.

The fact that Equation (11.2) takes the form

$$\text{logit probability} \sim \text{N(predicted value, variance)}$$

rather than

$$\text{logit probability} = \text{predicted value}$$

is important. It implies that we do not assume that the prior model formed by the β^0, β_c^{coun}, and β_t^{time} to perfectly predict $\text{logit}(\gamma_{ct})$ and hence γ_{ct}. Instead, we assume that the predicted values will sometimes be too high, and sometimes be too low. The size of these errors is reflected in the variance term.

In Section 8.2 we saw how the posterior distribution for an unknown quantity is a compromise between the likelihood and the prior. The posterior distribution for a probability γ_{ct} is one such compromise. It is a weighted combination of the likelihood, represented by the direct estimate y_{ct}/w_{ct}, and the prior model, represented by the predicted value $\tilde{\gamma}_{ct}$. The relative weights given to the likelihood and prior depend on the amount of data available for cell ct. The more data there are for a cell, the more weight is placed on the likelihood.

likelihood   prior

$\longleftarrow$ ——————————————————————————————— $\longrightarrow$

more data   less data

This means that the likelihood-based measure, the direct estimate y_{ct}/w_{ct}, receives the most weight in the cases when it works best, that is, when the data are most abundant. Conversely, the prior-based measure $\tilde{\gamma}_{ct}$ receives more weight in the cases where the direct estimate performs badly, or when the data are scarce. In intermediate cases, where there is a moderate amount of data, likelihood and prior contribute more evenly.

11.3.3   Prior for Region Effect

Next we add some substance to the idea that rates vary by county. We do this by specifying a prior for the county effect. The prior we use is a simple exchangeable prior stating that county effects are clustered around a central value,

$$\beta_c^{\text{coun}} \sim \text{N}(0, \tau_{\text{coun}}^2).$$

We set this central value to 0. It turns out that the choice of central value makes little difference to any of the results, other than the value of β^0.

The value of τ_{coun}, in contrast, does matter. Smaller values for τ_{coun} are associated with tighter clustering of the county effects. Tighter clustering of county effects implies less geographical variation in the γ_{ct}. If our estimates of geographical variation are to be correct, it is important that we choose an appropriate prior distribution for τ_{coun}.

We take the same approach to τ_{coun} that we do with most parameters governing variation: we use a weakly informative prior (discussed in Section 8.8.1). A weakly informative prior for a parameter such as τ_{coun} rules out implausibly large values, and mildly favours small values.

The prior we use is a half-t distribution with seven degrees of freedom and scale 1,

$$\tau_{\text{coun}} \sim t_7^+(1).$$

This prior is graphed in Figure 8.9. It places almost no weight on values larger than 4, and substantial weight on values less than 1.

Some insights into τ_{coun} and its prior can be obtained through a few back-of-the-envelope calculations. When τ_{coun} equals 1, region effects follow a normal distribution with mean 0 and standard deviation 1. About 95% of draws from a normal distribution with mean 0 and standard deviation 1 can be expected to fall between -2 and 2. A difference of 4 points on a logit scale is equivalent to moving from just above 0 to 0.1 on a probability scale. In practice we would not expect regional infant mortality rates in a country like Sweden vary by such a large amount. In other words, a half-t prior with scale 1 implies more variability in regional infant mortality rates than we would genuinely expect. The prior pulls the estimate of τ_{coun} away from large values, but the effect is weaker than it would be if the prior fully reflected our beliefs. This allows estimates of the region effects to be guided more by the data than by the prior.

11.3.4 Prior for Time Effect

The local level model introduced in Section 8.5.3 is often a good choice for modeling change over time. However, it lacks any way of representing sustained trends upwards or downwards. As can be seen in Figure 11.2, infant mortality rates in Sweden have a clear downward trend. The local level model is therefore not the best choice for these data.

A better choice is an extension of the local level model known as the local trend model. This model is depicted in Figure 11.4. The first layer of a local trend model is identical to a local level model. Each unit (in Figure 11.4, the β_t^{time}) equals a level term α_t plus some random error,

$$\beta_t^{\text{time}} \sim N(\alpha_t, \tau_{\text{time}}^2). \tag{11.3}$$

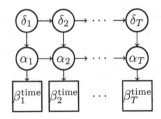

FIGURE 11.4: A local trend model. The α_t describe the level, and the δ_t describe the trend.

The second layer, however, differs between the two models. In the local level model, the level term α_t equals the previous level term α_{t-1} plus random error. In the local trend model, the level term equals the previous level term α_{t-1} plus a trend term δ_t plus random error,

$$\alpha_t \sim N(\alpha_{t-1} + \delta_t, \omega_\alpha^2). \tag{11.4}$$

If the trend term is negative, then the series is trending downwards; if the trend term is positive, then the series is trending upwards. Large values for δ_t imply rapid change.

The direction and size of the trend is not, however, fixed. Instead it evolves across units. The evolution is modeled using a random walk,

$$\delta_t \sim N(\delta_{t-1}, \omega_\delta^2). \tag{11.5}$$

(See Section 8.5.3 for a discussion of random walk models.)

The three sorts of random errors capture different aspects of the time effects. The random error in β_t^{time} only affects the time effect for year t. The random error in α_t permanently changes the mean of time effects for subsequent years. The random error in δ_t permanently changes the mean of year-on-year differences in time effects for subsequent years.

For each of the standard deviations for the random errors (τ_{time}, ω_α, and ω_δ), we use a weakly informative prior, a half-t distribution with seven degrees of freedom and scale 1.

11.3.5 Prior for Intercept

The intercept β^0 is a single parameter, and affects every observation. This makes it easy to estimate. Any reasonable prior for β^0 is overwhelmed by the data. We do not bother to specify a complicated or informative prior, but instead use a simple, weak one:

$$\beta^0 \sim N(0, 10^2). \tag{11.6}$$

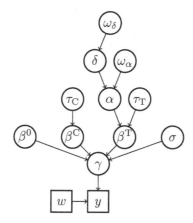

FIGURE 11.5: Summary of model for Swedish infant mortality rates. y denotes infant deaths, and w births. γ is infant mortality rates. β^{C} (shortened from β^{coun} in the text) is region effects, and β^{T} (shortened from β^{time}) is time effects. α is the level and δ the trend in the prior for time effects. The intercept β^0 has a normal prior with standard deviation 10. Standard deviation terms σ, τ_{C} (shortened from τ_{coun} in the text), τ_{T} (shortened from τ_{time} in the text), ω_α and ω_δ all have half-t priors with 7 degrees of freedom and scale 1.

11.3.6 Prior for Standard Deviation

The standard deviation term σ, from Equation (11.2), like all the other standard deviation terms, receives a half-t distribution with 7 degrees of freedom and scale 1.

11.3.7 Summary

Figure 11.5 summarizes the model as a whole. The only quantities in the model that are observed directly are deaths y and births w. Everything else must be estimated.

11.4 Results

11.4.1 Infant Mortality Rates

Figure 11.6 shows estimates of the underlying infant mortality rates. The credible intervals are wider, implying greater uncertainty, in counties with small populations than in counties with large populations. This makes sense. There is greater random variability in death counts in counties with small populations, and hence less information on which to base the estimates.

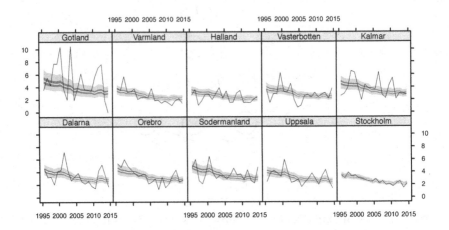

FIGURE 11.6: Estimates of infant mortality rates from the model given in Section 11.3. The light gray bands show 95% credible intervals, and the dark gray bands show 50% credible intervals. The white lines are posterior medians. The black lines are direct estimates.

The estimates lie closer to the prior model—that is, they smooth more—when data are limited. Between 2004 and 2005, for instance, the direct estimate for Gotland fell from 10.5 per thousand to 1.9 per thousand. These numbers are, however, based on deaths counts of 5 and 1, which are too small to support reliable estimates. Accordingly, the final estimates lean more heavily on the rates predicted by the prior model, and Figure 11.6 shows only a small dip between 2004 and 2005.

Overall, the model seems to do a good job of smoothing through the annual fluctuations to reveal underlying trends. There are no particular regions or years where the modeled rates seem to depart systematically from the direct estimates.

11.4.2 Intercept, Region Effects, and Time Effects

Figure 11.7 shows results for the intercept, region effects, and time effects. The graphs use different scales.

The credible interval for the intercept is quite narrow, implying that the intercept was estimated precisely. The posterior median is -5.48 on the logit scale. A value of -5.48 on the logit scale maps to 0.004 on a probability scale, which is close to the average infant mortality rate for the whole period. Intercepts generally represent some sort of average, so this close relationship is to be expected.

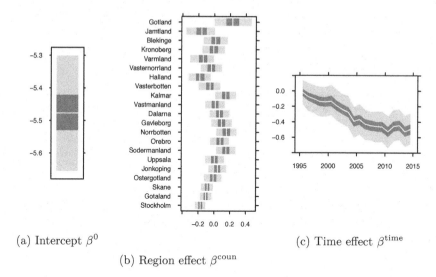

(a) Intercept β^0

(c) Time effect β^{time}

(b) Region effect β^{coun}

FIGURE 11.7: Estimates of the intercept, and region and time effects, from the model in Section 11.3. All estimates are on the logit scale. The light gray bands represent 95% credible intervals, the dark gray bands represent 50% credible intervals, and the white lines represent posterior medians.

The counties in the graph of region effects are ordered from top to bottom by increasing population size. Region effects are estimated more precisely for large regions than for small ones. There is evidently some regional variation in rates. Region effects for counties such as Halland and Stockholm are notably lower than region effects for counties such as Gotland and Kalmar.

The time effects fluctuate from year to year. There is, however, a clear downward trend in the earlier years, though this appears to flatten out towards the end of the period.

11.4.3 Prior for Time Effect

Results for the local trend model for the time effect are shown in Figure 11.8. The level term is basically the time effect with the annual fluctuations smoothed away. Removing the annual fluctuations, makes the slow-down in mortality decline more apparent. The trend term starts in the neighbourhood of -0.03, and by the end of the period is about half that size.

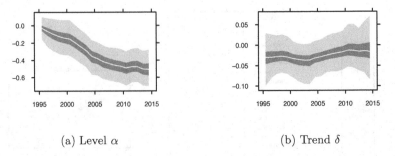

(a) Level α (b) Trend δ

FIGURE 11.8: Estimates of level and trend terms from the prior for the time effect. All estimates are on a logit scale, with a different scale used in each graph. The light gray bands represent 95% credible intervals, the dark gray bands represent 50% credible intervals, and the white lines represent posterior medians.

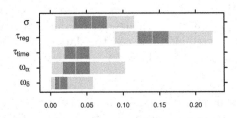

FIGURE 11.9: Estimates of standard deviation terms.

11.4.4 Standard Deviations

The last set of estimates are the standard deviation terms, shown in Figure 11.9. The prior for the region effect contains only one standard deviation term, while the prior for the time effect contains three.

11.5 Model Checking

The results presented so far seem reasonable. However, as we discuss in Section 9.6, it is unwise to place too much faith in results from a model until the model has been subjected to model checking.

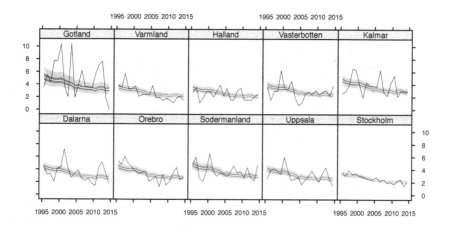

FIGURE 11.10: Direct estimates for infant mortality rates versus predictions from prior model. The direct estimates are defined in Section 11.2, and the predictions are defined in Section 11.3.2. The light gray bands show 95% credible intervals, and the dark gray bands show 50% credible intervals. The white lines are posterior medians. The black lines are direct estimates.

11.5.1   Model Predictions versus Direct Estimates

A useful general-purpose technique for checking hierarchical models is to compare direct estimates with the values predicted by the prior model. In the Swedish infant mortality case, we compare direct estimates y_{ct}/w_{ct} with the predicted values $\tilde{\gamma}_{ct}$ defined in Section 11.3.2. In contrast to the γ_{ct} from the lowest level of the model, which adapt to local features of the data, the $\tilde{\gamma}_{ct}$ represent the prior model alone. If the prior model is missing some important details, comparison with the direct estimates may reveal it.

Getting the prior model right is important in applications where there are missing data, since estimates for the cells with missing data are based solely on the prior model. This includes forecasting, where the missing data are for future periods. (See Section 9.5 for discussion of this point.)

The direct estimates and predicted values are compared in Figure 11.10. It turns out that the predicted values are close to the rates shown in Figure 11.6. The plot gives no obvious cause for concern, though results for small counties are difficult to interpret, because the direct estimates are so erratic.

11.5.2   Regional Variation in Slopes

By including a regional effect in the prior model, we allow for the possibility that the average level of mortality differs across counties. However, by using a single time trend for all regions, we are saying that we expect mortality in all

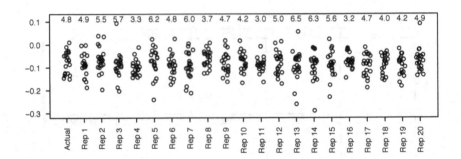

FIGURE 11.11: Using replicate data to check the assumption that infant mortality rates decline at the same pace in all counties. Each circle shows the slope from a linear regression fitted through the direct estimates for a county. The first set of points were calculated from the actual data. The remaining 20 sets were calculated from 20 replicate datasets. The numbers at the top are the slopes, multiplied by 1,000.

counties to decline at the same rate. The prior model does not allow for the possibility that some counties experience more rapid mortality decline than others, except for temporary deviations from trend (captured by the random error in $\text{logit}(\gamma_{ct})$).

As can be seen in Figure 11.6 or Figure 11.10, there are some counties, such as Halland, where the direct estimates appear to decline more slowly than the modeled estimates, and other counties, such as Varmland, where they appear to decline more quickly. Given the variability in the direct estimates, however, it is difficult to know whether this constitutes strong evidence against the assumption of uniform rates of decline.

We investigate further, using the replicate data techniques introduced in Section 9.6.3. We compare the variation across counties in the rate of the decline for the real dataset with variation across counties in 20 replicate datasets. If variation in the real dataset was greater than variation in the 20 replicate datasets, it would suggest that the model is deficient, and that it is inappropriate to assume a common rate of decline.

To generate the rate of decline for one set of replicate data, we proceed as follows:

1. Use Equation (11.2), plus draws from the posterior distributions for the intercept, region effect, and time effect, to randomly generate predicted mortality rates.

2. Use Equation (11.1) to randomly generate death counts.

3. Use these death counts and the observed births to calculate direct estimates of infant mortality rates.

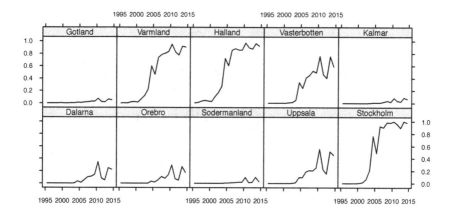

FIGURE 11.12: Probability that the underlying infant mortality rate is less than 2.5 per thousand

    4. For each county, fit a straight line through the direct estimates, and record the slope of the line.

Each repetition of this procedure yields a set of 21 slopes, for the 21 counties of Sweden. Fitting straight lines through the actual direct estimates yields a further set of 21 slopes.

The results are shown in Figure 11.11. The actual data fall somewhere in the middle of the range for variability. These results suggest, in assuming a common trend in the prior model, we have not over-simplified.

11.6   Summarizing Results via Probabilities

As we saw in Section 9.2, a sample from the posterior distribution can be used to produce summaries that are customized to the application at hand. To illustrate, we calculate the probability that the underlying infant mortality rate in each county in each year is less than 2.5 per thousand.

If we have a sample from the posterior distribution, then estimating the probability that a county's infant mortality rate is less than 2.5 per thousand is easy. We simply count the number of draws from the sample in which the mortality rate is below 2.5 per thousand, and divide by the total number of draws. In Gotland in 1995, for instance, none of the draws is below 2.5 per thousand, so we set the probability to 0. In Varmland in 2004, 21% of the draws is below 2.5 per thousand, so we set the probability to 0.21.

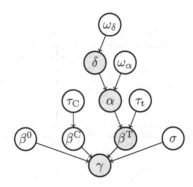

FIGURE 11.13: Forecasting Swedish infant mortality rates. See the key for Figure 11.5 for a description of the components. The gray components are time-varying. Forecasted values can be generated by generating values for the time-varying components, and combining them with existing values for the time-invariant components.

The measure is a little noisy, as evident from the spikes and troughs in Figure 11.12. However, it does provide a succinct summary of the progress made by each county.

11.7 Forecasting

Next we forecast underlying infant mortality rates, by county, for the years 2016–2025. We start by outlining the process for constructing a forecast.

11.7.1 Constructing the Forecasts

When we have a sample from the posterior distribution for the past rates and parameters in the prior model, constructing a sample from the posterior distribution for the future rates is easy. The process is summarized in Figure 11.13. We already have values for parameters that do not vary over time, such as the region effects and the standard deviations ω_δ and ω_α. Generating forecasts therefore reduces to generating values for the time-varying components, and combining them with the time-invariant components.

To generate a single draw from the posterior distribution for the forecasts, we go through the following steps:

1. Plug values for ω_δ into Equation (11.5), and generate values for δ.

2. Plug the values for δ and ω_α into Equation (11.4), and generate values for α.

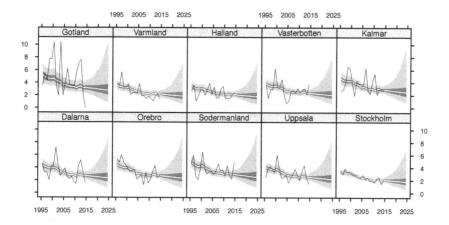

FIGURE 11.14: Estimates and forecasts for underlying infant mortality rates in Sweden. The light gray bands represent 95% credible intervals, the dark gray bands represent 50% credible intervals, and the white lines represent posterior medians. The black lines represent direct estimates.

3. Plug the values for α and τ_{time} into Equation (11.3), and generate values for β^{time}.

4. Plug values for β^{time}, β^0, β^{coun}, and σ into Equation (11.2), and generate values for γ_{ct}.

The forecasts are generated from exactly the same model that is used to generate the estimates. The only difference is that the forecasts do not involve the likelihood (Equation (11.1)), for the simple reason that there are no data for future periods. Instead, the forecasts rely entirely on the prior model, as fitted to the historical data.

11.7.2 Results: Exploding Credible Intervals for Forecasting

The results from the forecasting are summarized in Figure 11.14. The 50% credible intervals, and the posterior medians, seem fine. The 95% credible intervals, however, do not. The upper bounds imply that there is a 2.5% probability that underlying infant mortality rates will be about twice as high in 2025 as they were in 1995. This is not plausible.

The phenomenon of exploding credible intervals is, unfortunately, common in forecasting. More precisely, it is common for posterior medians and 50% credible intervals to conform to patterns in the historical data, and to prior expectations, but for more comprehensive credible intervals, such as 90% or 95% credible intervals, to cover an implausibly wide range.

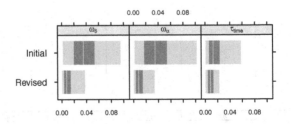

FIGURE 11.15: Posterior distributions for standard deviation terms with initial weak priors and revised stronger priors.

The wide 95% credible intervals in Figure 11.14 reflect the fact that some values in the sample from the posterior distribution for the γ_{ct} are very high or very low. Values of γ_{ct} are typically extreme when the standard deviation terms in the prior for the time effect, ω_δ, ω_α, and τ_{time}, are large. When these values are large, the time effect, and hence prior value $\tilde{\gamma}_{ct}$, and hence forecasted probability γ_{ct} can take extreme values. Large values for ω_δ and ω_α have a particularly strong impact, since their effect is cumulative. When ω_δ and ω_α are large, the trend and level terms can take a succession of large jumps, and end up far away from their expected values.

In a mathematical sense, then, the exploding credible intervals in Figure 11.14 reflect the fact that the posterior distributions for standard deviation terms ω_δ, ω_α, and τ_{time} include some large values. As can be seen in Figure 11.9, the 95% credible interval for ω_α extends up to 0.1. This implies that we would frequently see annual shifts in the underlying infant mortality rate of 10%, 20% or more. We certainly do not see such shifts. What has gone wrong?

The problem is that we have a weak prior and weak data. Our prior for the standard deviations has a scale of 1, which does nothing to rule out values like 0.1. Meanwhile, the model only has 21 time points with which to estimate the three standard deviation terms. With hindsight, it is not surprising that the posterior distribution is wider than it ought to be.

11.7.3   A Partial Solution

One potential solution to the problem is to strengthen the priors for standard deviation terms ω_δ, ω_α, and τ_{time}. To demonstrate, we set the scale parameters in the priors for these terms to 0.01. The value 0.01 is chosen on the grounds that infant mortality in Sweden should vary from year to year by a few percentage points at most. Three error terms, each with standard deviations of 0.01, should produce variation of roughly this amount.

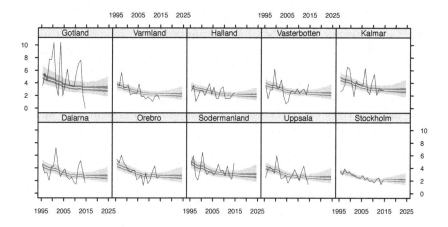

FIGURE 11.16: Revised estimates and forecasts for infant mortality rates in Sweden. The light gray bands represent 95% credible intervals, the dark gray bands represent 50% credible intervals, and the white lines represent posterior medians. The black lines represent direct estimates.

Figure 11.15 shows posterior distributions for the standard deviation terms based on the original weak priors and on the revised stronger priors. As we would expect, the posterior distributions based on the stronger priors are much more strongly concentrated on values close to 0.

The smaller standard deviations translate into narrower credible intervals, as can be seen in Figure 11.16. The revised forecasts still allow for the possibility of mortality rates rising over time, but give sharp rises a much lower probability than before.

Our solution to the problem of exploding credible intervals for forecasting is, however, only partial. The use of 0.01 as a scale for the priors seems broadly reasonable, and leads to more believable forecasts than the original specification. We return to the topic of choosing model specifications in the next chapter.

11.8 References and Further Reading

The data on births and deaths in Sweden were obtained from the tables "Live births by region, sex and age of mother. Year 1968–2015" and "Deaths by region, age (during the year) and sex. Year 1968–2015", downloaded on February

17-18, 2017. The data can be found on the website for the book, www.bdef-book.com.

UN Inter-agency Group for Child Mortality Estimation (2013, 2014, 2015, 2017) uses Bayesian methods to produce estimates of infant mortality rates for countries with high-quality vital registration data.

Gelman and Hill (2007) discuss how the posterior distribution is a compromise between direct estimates and the prior model. Prado and West (2010, ch. 4) describe the local trend model. Gelman (2006) and Gelman et al. (2008) make the case for using a half-t prior as a default prior distribution for standard deviation parameters.

Bryant and Zhang (2016) develop a hierarchical model for migration similar in spirit to the one developed in this chapter, submit the model to validity tests, and use it for forecasting.

12

Life Expectancy in Portugal

In this chapter, we continue modeling mortality, but with a change in country and a change in dimensions. The new country is Portugal, and dimensions are age, sex, and time.

A major theme of this chapter is interactions between age, sex, and time. Variables interact when the nature of the relationship between a variable and the outcome of interest depends on the level of one or more other variables. If, for instance, the relationship between age and mortality differs between females and males, then we say that there is an interaction between age and sex. In most demographic datasets, there are interactions that require attention from modelers. In this chapter, we pay particular attention to modeling how age patterns and sex differences change over time.

Another theme of the chapter is model comparison using heldback data. Comparing models using heldback data is a standard solution to the problem, discussed at the end of Chapter 11, of choosing a good model for forecasting. In this chapter, we held back data to choose a model to use for forecasting Portuguese life expectancy up to 2035.

12.1 Mortality Rates

Our deaths data consist of annual death counts and exposures, for age groups 0, 1–4, 5–9, $\cdots$, 94–99, 100+, by sex, for the period 1950–2015. Splitting the age group 0–4 into separate age groups 0 and 1–4 is common with mortality statistics. Mortality rates for infants tend to be much higher than for other children, and it is often helpful to deal with infants separately. The age group 100+ refers to everyone aged 100 and over. This group is usually small, because few people live this long.

In addition to the deaths data, we have data on the size of the population exposed to the risk of dying. The exposures are disaggregated in the same way as the death counts.

Figure 12.1 shows direct estimates of mortality rates, constructed by dividing the death counts by the exposures. The rates vary by several orders of magnitude. In 1950, for instance, they range from 0.00115 (for 10–14 year olds) to 0.362 (for age 100+). When values vary by as much as this, plot-

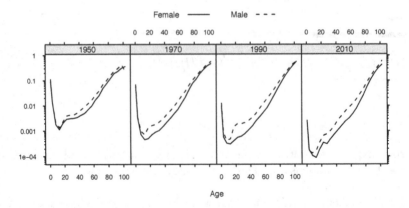

FIGURE 12.1: Direct estimates of age-specific mortality rates in Portugal, on a log scale. The rates are obtained by dividing deaths counts by exposures. The lines are drawn through the center of each age group.

ting them on an ordinary scale obscures differences among smaller values. We therefore plot them on a log scale.

12.2   Log Function

The log function, like the logit function discussed in Section 11.3.2, maps a restricted scale on to an unrestricted one. The log function maps points that must be greater than 0 to the log scale, where there are no upper or lower bounds.

The log scale emphasizes relative differences rather than absolute differences. On a log scale, for instance, the difference between 0.001 and 0.002 is equivalent to the difference between 1 and 2.

Figure 12.2 illustrates the effects of a log transformation, using direct estimates of mortality rates for Portugese females in 2015. Below age 80, the absolute differences between age groups are small, and the relative differences are moderate. Above age 80 the absolute differences are large, while the relative differences are again moderate. Panel (a) brings out the absolute differences, while Panel (b) brings out the relative differences.

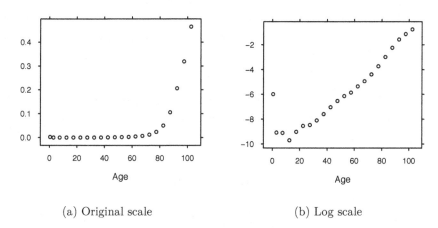

(a) Original scale   (b) Log scale

FIGURE 12.2: Direct estimates of mortality rates for Portugese females in 2015, on the original scale and the log scale. Note that the vertical axis of Panel (b) shows values after transformation to the log scale, whereas the vertical axis of Figure 12.1 shows values before transformation. Both types of annotation are common with the log scale.

The inverse of the log function is the exponential function with base e. Thus if

$$y = \log(x),$$

then

$$x = \exp(y).$$

A convenient feature of the log function is that, for small values of x, a difference of x on the log scale translates to a relative differences of $x \times 100\%$ on the original scale. For instance, suppose that, when measured on the log scale, the mortality rate for group A is 0.1 higher than the mortality rate for group B. If we convert back to the orginal scale, we will find that the resulting value for group A is almost exactly $0.1 \times 100\% = 10\%$ higher than the value for group B.

12.3   Life Expectancy

Life expectancy is the remaining years an individual could be expected to live, under a given set of mortality rates. Life expectancy is usually calculated for individuals aged exactly 0, but it can be calculated for individuals of any age. We can, for instance, calculate life expectancy at age 65, which gives the number of additional years an individual could expect to live from the day of their 65th birthday.

Demographers distinguish between "period" and "cohort" life expectancy. With period life expectancy, the mortality rates refer to a particular year, such as 2015. With cohort life expectancy, the mortality rates are those experienced by an actual cohort (Section 4.7). Cohort life expectancy for the cohort born in 2015, for instance, would use 2015 rates for age 0, 2016 rates for age 1, 2017 rates for age 2, and so on. In this book, when we refer to life expectancy, we mean period life expectancy. Formulas for calculating life expectancy can be found in most demographic textbooks.

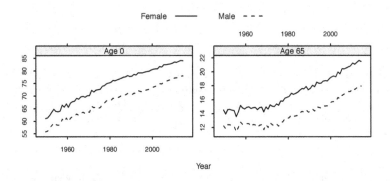

FIGURE 12.3: Direct estimates of life expectancy at ages 0 and 65 in Portugal.

Life expectancies for Portugal, calculated using the direct estimates of mortality rates, are shown in Figure 12.3. The left panel shows life expectancy at age 0, and the right panel shows life expectancy at age 65. As is typical for modern populations, female life expectancy is greater than male life expectancy at birth and in the older ages. Life expectancy at age 65, for females and males, showed no clear trend until about 1970, after which it climbed steadily upward.

12.4 Age, Sex, and Time Effects

Examining the mortality rates in Figure 12.1 by eye, it is easy to see, in broad terms, how mortality varies with age, sex, and time. In other words, it is easy to identify an age effect, a sex effect, and a time effect:

Age effect. Mortality is high at age 0 before dropping to very low levels. It then climbs steadily up to age 100+.

Sex effect. Mortality is lower, overall, for females than for males.

Time effect. Mortality rates have been trending downwards over time.

Sometimes, however, the effects are not so clear. It is therefore helpful to have a formal method for identifying possible effects. One such method is to 'decompose' direct estimates of the rates. As with Figure 12.1, we work with a logged version of the direct estimates. The age effect at age a describes how much the average log mortality rate for age a differs from the overall average log mortality rate. Because the age effects are all relative to the overall average, they sum to zero. Variability in age effects represents the contribution of age to overall variability of log mortality rates. Sex and time effects are defined similarly.

Let m_{ast} denote the direct estimate of log mortality rate for age group a, sex s, and year t, where $a = 1, \cdots, A$, $s = 1, 2$, and $t = 1, \cdots, T$. The overall average log mortality rate is then

$$\lambda_0 = \frac{1}{2AT} \sum_{a=1}^{A} \sum_{s=1}^{2} \sum_{t=1}^{T} m_{ast}, \tag{12.1}$$

the age effect is

$$\lambda_a^{\text{age}} = \frac{1}{2T} \sum_{s=1}^{2} \sum_{t=1}^{T} m_{ast} - \lambda_0, \tag{12.2}$$

the sex effect is

$$\lambda_s^{\text{sex}} = \frac{1}{AT} \sum_{a=1}^{A} \sum_{t=1}^{T} m_{ast} - \lambda_0, \tag{12.3}$$

and the time effect is

$$\lambda_t^{\text{time}} = \frac{1}{2A} \sum_{a=1}^{A} \sum_{s=1}^{2} m_{ast} - \lambda_0. \tag{12.4}$$

Results from the decomposition are shown in Figure 12.4. The decomposition confirms, for instance, that female rates are lower than males, and that mortality rates have been trending downwards over time. By looking at the vertical scales of the three graphs we also get a sense of the relative importance of the three effects. The estimated age effects are by far the most important of the three, ranging from approximately -2.5 to 4. The time effects, in turn, have about three times the range of the sex effects.

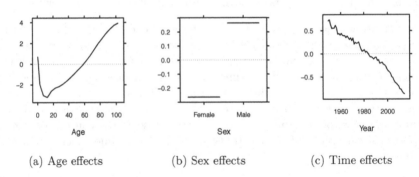

(a) Age effects   (b) Sex effects   (c) Time effects

FIGURE 12.4: Age, sex and time effects obtained by decomposing direct estimates of log mortality rates. Note that the three graphs use different vertical scales.

12.5 Interactions

Looking closely at Figure 12.1, it appears that female-male differences in mortality are not constant over the whole age range. Females have lower mortality above age 15 and below age 70, but not in the lowest or highest age groups. When describing the relationship between mortality, age, and sex, we can distinguish between an age effect, a sex effect, and an age-sex 'interaction'. The age-sex interaction captures the way that sex differences vary across the age range—or, equivalently, how age differences vary between sexes.

Similarly, looking at Figure 12.1, it appears that mortality rates have not declined at the same pace for all age groups. We can distinguish between an age effect, a time effect, and an age-time interaction. The age-time interaction captures the fact that mortality declines have been faster at younger ages than at older ages.

Interactions are common in demographic data, and often large enough that we need to include them in our models. Indeed, a substantial part of the process of building a demographic model consists of looking for interactions, and deciding how to handle them.

We can quantify interactions by extending the decomposition technique we used with age, sex, and time effects. Again, the basic strategy is to subtract away averages, and see what remains. For instance, the interaction effect for age a and sex s equals the average log mortality rate for age a and sex s, minus the overall average log mortality rate, minus the age effect for age a, and minus the sex effect for sex s.

The age-sex interaction is

$$\lambda_{as}^{\text{age:sex}} = \frac{1}{T} \sum_{t=1}^{T} m_{ast} - \lambda_0 - \lambda_a^{\text{age}} - \lambda_s^{\text{sex}}, \tag{12.5}$$

the age-time interaction is

$$\lambda_{at}^{\text{age:time}} = \frac{1}{2} \sum_{s=1}^{S} m_{ast} - \lambda_0 - \lambda_a^{\text{age}} - \lambda_t^{\text{time}}, \tag{12.6}$$

and the sex-time interaction is

$$\lambda_{st}^{\text{sex:time}} = \frac{1}{A} \sum_{a=1}^{A} m_{ast} - \lambda_0 - \lambda_s^{\text{sex}} - \lambda_t^{\text{time}}. \tag{12.7}$$

The results from applying these techniques to the Portugese data are graphed in Figure 12.5. Panel (a) shows the residual age profile for females, after accounting for the age effects, and the sex effect for females. The fact that the interaction is positive up to age 15 means that female rates are higher than expected at these ages, given the average age profile and the average sex effect for females.

The sex-time interaction in panel (b) is in interpreted in a similar way to the age-sex interaction. It measures the residual, after accounting for the sex and time main effects. (In statistical terminology, a 'main effect' is an effect involving only one dimension, such as an age effect or sex effect, as opposed to an interaction, which involves two or more dimensions.) The results in Figure 12.5b imply that female mortality rates were relatively high, compared with males, in 1950; that the gap narrowed until about 1980; and that the gap has remained stable since then. These results can be verified by careful inspection of the raw female and male age profiles in Figure 12.1.

Finally, the estimated age-time interaction in Figure 12.5c confirms that mortality rates have fallen faster for the young than the old, though the effect is most marked at the extreme ends of the age distribution.

We could extend the decomposition to estimate third-order-interactions between age, sex, and time. An age-sex-time interaction is in fact visible in the raw rates in Figure 12.1. Between 1970 and 1990, the age profile for males, but not females, develops a hump around age 20. Between 1990 and 2010, the hump disappears again. This hump around age 20 is a feature of mortality rates in many countries, and is known as the accident hump. To describe this phenomenon, a third-order interaction would be needed. But we will instead move on to model-building.

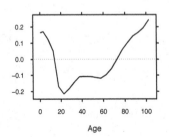

(a) Age-sex interaction: Females

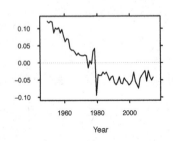

(b) Sex-time interaction: Females

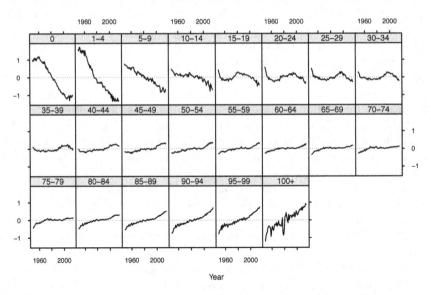

(c) Age-time interaction

FIGURE 12.5: Interactions obtained by decomposing direct estimates of log mortality rates. The vertical scales differ between graphs. The age-sex and sex-time interactions are for females.

12.6 Models

12.6.1 Likelihood

Our ultimate objective is to forecast life expectancy for the years 2016-2035. To do this, we estimate and forecast γ_{ast}, super-population mortality rates by age, sex, and time. The input data are deaths y_{ast} and exposures w_{ast}. We treat deaths as draws from a Poisson distribution,

$$y_{ast} \sim \text{Poisson}(\gamma_{ast} w_{ast}). \tag{12.8}$$

12.6.2 Model for Mortality Rates

Our baseline model for (log) mortality rates is

$$\log \gamma_{ast} \sim \text{N}(\beta^0 + \beta_a^{\text{age}} + \beta_s^{\text{sex}} + \beta_t^{\text{time}} + \beta_{as}^{\text{age:sex}} + \beta_{st}^{\text{sex:time}} + \beta_{at}^{\text{age:time}}, \sigma^2). \tag{12.9}$$

Using log rates rather than rates has similar advantages to using logit probabilities rather than probabilities, as described in Section 11.3.2. After applying the transformation, we no longer have to worry about avoiding negative values.

Equation (12.9) contains all three main effects and all three second-order interactions, but not the third-order interaction between age, sex, and time. Instead, any variation left after accounting for main effects and second-order interactions is modeled as random (normally-distributed) noise.

Why omit the third-order interaction when we know that the data contains at least one feature—the accident hump—that requires this interaction? There is an important practical obstacle to including the third-order interaction: when we try to estimate the model, the MCMC simulation struggles to converge. Poor convergence is, however, often a symptom of deeper problems with the model itself. In this case it is probably a warning sign that the model is "too complicated".

Capturing what applied statisticians mean by "too complicated" is tricky. The basic idea, though, is that models that try to capture every feature of the data tend to also capture random, meaningless patterns. Then, when used for forecasting, the overly-complicated models project these random patterns into the future. Such forecasts tend to be poor.

It is also important to remember that Equation (12.9) is only a prior. A sufficiently strong signal in the data will pull the posterior away from the prior. We would therefore expect historical estimates produced by the model to include the accident hump, even if the prior does not.

In addition to the baseline model, we specify an alternative, simpler model. The alternative model is identical to that of Equation (12.9), except that it omits the sex-time interaction. We omitted the sex-time interaction because, judging by Figure 12.5 it is the smallest in magnitude of the three second-order interactions. We use the alternative model to demonstrate the process of choosing models using heldback data.

12.6.3   Prior for Age Effect

We use a local trend model (Section 11.3.4) as the prior for age effects. The local trend model was originally developed for modeling change over time, but it works equally well for modeling change over age. Age effects, like time effects, often have persistent trends upwards or downwards.

A particularly nice feature of using the local trend model to deal with mortality rates is that it generalizes the Gompertz model. The Gompertz model, published by Benjamin Gompertz in 1825, states that log mortality rates at older age groups increase linearly with age. Nearly 200 years of evidence have confirmed that this is often the case. The increasing trend, however, is "local" to older age groups. For the youngest age groups, there is typically a "local" decreasing trend.

We make one modification to the standard local trend model. We add a special covariate to the first level of the model that applies only to infants,

$$\beta_a^{\text{age}} \sim \begin{cases} N(\alpha_a + \psi, \tau^2) & \text{if } a \text{ refers to infants} \\ N(\alpha_a, \tau^2) & \text{otherwise.} \end{cases} \tag{12.10}$$

The ψ in Equation (12.10) measures the extra mortality experienced by infants, compared with what would be expected from the mortality of other children. By adding a covariate for infants, we are saying, in effect, that we do not regard the increment between infants and 1–4 year olds as exchangeable with the increment between other age groups (Section 8.5.3).

The other components of the prior remain the same as in the standard local trend model:

$$\alpha_a \sim N(\alpha_{a-1} + \delta_a, \omega_\alpha^2), \tag{12.11}$$

$$\delta_a \sim N(\delta_{a-1}, \omega_\delta^2). \tag{12.12}$$

(In Equations (12.10), (12.11), and (12.12), and in all other prior specifications in this chapter, level and trend terms, and standard deviations such as τ and ω_α, are specific to the prior in question. We have omitted reference to terms such as "age" in superscripts and subscripts in order to reduce clutter.)

We use our standard weakly informative half-t prior, with seven degrees of freedom and scale 1, for ψ. As usual, a weakly informative prior helps rule out implausible values, and speeds up computations.

12.6.4   Prior for Time Effect

For time effects, we again use the local trend model. However, we simplify the model slightly. Instead of assuming

$$\alpha_t \sim N(\alpha_{t-1} + \delta_t, \omega_\alpha^2), \tag{12.13}$$

we assume

$$\alpha_t = \alpha_{t-1} + \delta_t. \tag{12.14}$$

We remove the possibility of random shifts in the level term, and instead make the level completely dependent on the trend. The other components of the simplified prior remain the same as in the standard local trend model:

$$\beta_t^{\text{time}} \sim \text{N}(\alpha_t, \tau^2), \tag{12.15}$$

$$\delta_t \sim \text{N}(\delta_{t-1}, \omega_\delta^2). \tag{12.16}$$

We have found that, with the Portugese data, and with mortality data more generally, the simplified version of the local trend model tends to outperform the full version. We suspect that this is because discrete, permanent changes in the level of mortality, of the type modeled by random shifts in the level term, are relatively rare. Instead, long-term mortality levels tend to change slowly and continuously, in response to slow, continuous change in the determinants of long-term trends such as technology, institutions, and environmental conditions.

12.6.5 Prior for Age-Time Interaction

The prior for the age-time interaction is designed to capture the patterns that are apparent in the decomposition results in Figure 12.5c. Each age group a has its own time series. Each of these time series follows a simplified local trend model,

$$\beta_{at}^{\text{age:time}} \sim \text{N}(\alpha_{at}, \tau^2), \tag{12.17}$$

$$\alpha_{at} = \alpha_{a,t-1} + \delta_{at}, \tag{12.18}$$

$$\delta_{at} \sim \text{N}(\phi\delta_{a,t-1}, \omega_\delta^2). \tag{12.19}$$

The local trend model of Equations (12.17)–(12.19) differs from previous local trend models we have considered, however, in that it includes a damping term ϕ.

The presence of the damping term means that, instead of an ordinary random walk, trend terms δ_{at} follows a damped random walk. A damped random walk differs from an ordinary random walk in that each step tends to be smaller than the one before it. The decline in step size is governed by ϕ, which takes a value between 0 and 1. When ϕ is close to 0, step size declines quickly, and when ϕ is close to 1, it declines slowly. When ϕ equals 1 exactly, there is no damping, and the model reverts to a ordinary random walk.

Consider a damped local trend model where the trend term currently has a value of d, and the damping parameter has a value of 0.95. The expected value for the trend term next period is $0.95 \times d$; the expected value the period after next is $0.95 \times 0.95 \times d = 0.9025 \times d$; and so on. The expected value for the trend after 5 periods is 23% lower than the current value. Table 12.1 shows expected values for other choices of damping parameter and time period.

Damped local trend models are based on the principle that no upward or downward trend continues indefinitely. For most time series, this principle

TABLE 12.1
Percent reduction in expected value for trend term δ_{at} after 5,
10, and 20 years, for selected values of damping parameter ϕ

ϕ	Years		
	5	**10**	**20**
1.00	0	0	0
0.99	5	10	18
0.95	23	40	64
0.90	41	65	88
0.80	67	89	99

is borne out by the evidence. Empirical studies of the performance of time
series models generally find that models where trends are damped give more
accurate forecasts than models where trends are not damped.

In the particular case of age-time interactions for mortality rates, damping seems to us to be particularly appropriate. Human mortality rates have a
characteristic age-profile, which recurs, with variations, across many populations. Damping prevents forecasted age-profiles from departing too far from
their observed historical average, which, arguably, increases their plausibility.

The prior for damping parameter ϕ is depicted in Figure 12.6. This prior
restricts ϕ to lie between 0.8 and 1, while down-weighting values at the extreme
ends of this range. This prior is our Bayesian re-interpretation of the default
for the damping parameter in function `ets` in R package `forecast`. (The
package **forecasting** is one of the most popular packages for R, and has been
used for a huge variety of problems, which makes it a good place to look for
default settings.)

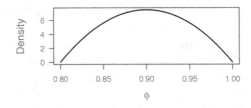

FIGURE 12.6: Prior for damping parameter ϕ. The prior is a beta distribution
with shape parameters 2 and 2, rescaled to lie on the interval between 0.8 and 1.

The beta distribution is a continuous probability distribution restricted to the range between 0 and 1. The beta distribution has two positive shape parameters that control the shape of the distribution. The probability density function for a beta distribution with shape parameters α and β is

$$p(y) = \frac{\Gamma(\alpha + \beta)}{\Gamma(\alpha)\Gamma(\beta)} y^{\alpha-1}(1 - y)^{\beta-1}.$$

The time series for different age groups all share the same values for τ, ω_δ, and ϕ. Estimates for the series therefore pool information across age groups. Ideally, they would pool even more information than they currently do. For instance, we might obtain more precise estimates if we took account of the fact that neighbouring age groups tend to follow similar trends. Extending the model in this way, however, would make the model overly-complicated and the estimation much more difficult.

12.6.6 Prior for Sex-Time Interaction

In the baseline model, the sex-time interaction receives the same sort of prior as the age-time interaction: the series are represented by a simplified, damped local trend model. As noted above, the alternative model does not include a sex-time interaction. It therefore omits this prior.

12.6.7 Priors for Other Terms

With the intercept and sex effect, we use simple priors

$$\beta^0 \sim N(0, 10^2) \tag{12.20}$$
$$\beta_s^{\text{sex}} \sim N(0, 1). \tag{12.21}$$

We use a local level model for the age-sex interaction. We experimented with a local trend model, but found that it gave less precise estimates, which suggests that the extra complexity is not warranted. The standard deviation terms all receive our usual weakly informative half-t prior.

12.6.8 Summary

Figure 12.7 summarizes baseline and alternative models, focusing on terms governing the means. As is apparent from all the superscripts that contain "T" (for "time"), both models provide plenty of scope for demographic patterns to change over time. The baseline model, however, provides a little more flexibility than the alternative model.

We will use heldback data to compare models, and see if the extra flexibility improves forecasts. Before doing so, however, we briefly summarize the key ideas behind the use of heldback data.

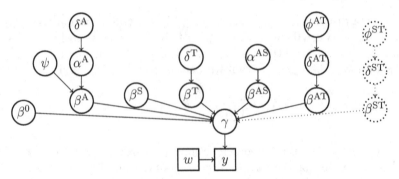

FIGURE 12.7: Summary of models. In the superscripts, "age" is abbreviated to "A", "sex" to "S", and "time" to "T". For simplicity, standard deviations have been omitted. The sex-time interaction is present in the baseline model but not the alternative model.

12.7   Model Choice Using Heldback Data

The steps involved in using heldback data to choose a forecasting model are depicted in Figure 12.8. The full dataset is partitioned into a "training" dataset and a "test" or "heldback" dataset. Two or more models are then fit to the training dataset. The fitted models are used to obtain forecasts for the period covered by the test dataset. The test dataset is then used to assess the performance of each forecast. The model with the best performance wins.

12.8   Estimating and Forecasting

We partition the Portugese mortality data into a training dataset extending from 1950 to 1995 and a test dataset extending from 1996 to 2015. We fit the baseline model to the training dataset and then do the same with the alternative model. Then, with each model, we forecast the deaths rates γ_{ast} over the period 1996–2015.

The process of forecasting the γ_{ast} is summarized in Figure 12.9. The gray quantities are time-varying, and everything else is time-invariant. The time-invariant quantities are fed into the forecast as they are, but values for the time-varying quantities must be generated from the model. This is done by starting with the parameters at the top of the model, and working downwards. In the baseline model, for instance, we generate new values for time trends δ^{T}, δ^{AT}, and δ^{ST}, then new values for β^{T}, β^{AT}, and β^{ST}, then new values for γ. More details are given in Section 12.12.

(1) Partition the data into "training" and "test" datasets

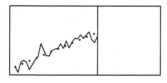

(2) Fit both models to the training dataset

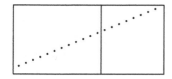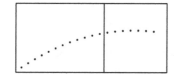

(3) Use the fitted models to make forecasts

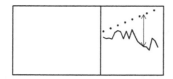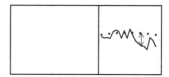

(4) Use the test dataset to evaluate the forecasts

FIGURE 12.8: Comparing two models using heldback data. The model on the right out-performs the model on the left.

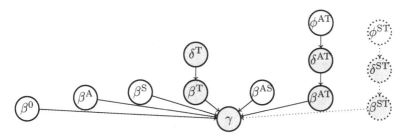

FIGURE 12.9: Forecasting mortality rates. Time-varying quantities are shaded gray. The sex-time interaction is present in the baseline model but not the alternative model.

12.9   Comparing the Forecasts with the Heldback Data

Comparing forecasts with heldback data is often more complicated than it first appears. In our case, it involves the following steps:

1. Calculate observed mortality rates from the heldback data.
2. Convert the forecasted γ_{ast} into forecasted finite-population rates.
3. Summarize the observed and forecasted rates using life expectancies.
4. Using a variety of performance measures to assess the agreement between the forecasted and observed life expectancies.

Step 1 is straightforward. We simply divide the heldback deaths y_{ast} by the heldback exposures w_{ast}.

The reason for Step 2 is rather subtle. The forecasted γ_{ast} are super-population quantities, while the observed rates are finite-population quantities. As we discuss in Section 4.9, finite-population quantities tend to be more variable then their super-population equivalents, since finite-population quantities are affected by the random variation in the number of events. When cell sizes are all large, this extra variation can be safely ignored, but when sizes are small, it can be important to include it. We convert the forecasted γ_{ast} into finite-population rates by using Equation (12.8) to randomly generate death counts, and then dividing these counts by observed exposures.

Life expectancy is a complicated, non-linear function of age-specific mortality rates. As we discuss in Section 9.3.1, however, Step 3 is, nevertheless, easy.

To perform Step 4, we need to decide on some performance measures. We choose three: (i) absolute error, (ii) interval score, and (iii) continuous ranked probability score (CRPS). For all three performance measures, smaller values imply better performance.

Absolute error measures the performance of point estimates generated by a forecast. We use posterior medians of life expectancies as our point estimate. The absolute errors are the absolute differences between point estimates and observed values.

Let $\hat{e}_{ast}$ denote the posterior median of life expectancy at age a for sex s and year t. Let e_{ast} denote the observed life expectancy at age a for sex s and year t. The absolute error is

$$|\hat{e}_{ast} - e_{ast}|.$$

The interval score measures the performance of a particular size of credible interval, e.g. an 80% interval, produced by the model. The interval score rewards narrow intervals, but also penalizes intervals that do not contain the observed value.

For life expectancy at age a for sex s and year t, let $\hat{Q}_{ast}(p)$ denote the $100p\%$ quantile of the posterior distribution, where p is a given probability. The lower and upper bounds of the $(1-\alpha) \times 100\%$ credible interval are given by $\hat{A}_{ast} = \hat{Q}_{ast}(\alpha/2)$ and $\hat{B}_{ast} = \hat{Q}_{ast}(1-\alpha/2)$. The interval score is

$$(\hat{B}_{ast} - \hat{A}_{ast}) + \frac{2}{\alpha}(\hat{A}_{ast} - e_{ast})I\left(e_{ast} < \hat{A}_{ast}\right)$$
$$+ \frac{2}{\alpha}(e_{ast} - \hat{B}_{ast})I\left(e_{ast} > \hat{B}_{ast}\right).$$

Here $I(\cdot)$ is an indicator function, taking value 1 if the condition given in the parentheses is satisfied, and 0 otherwise.

The first term in the interval score equals to the length of the credible interval. The second term gives a penalty if the observed value is smaller than the lower bound of the credible interval. The third term gives a penalty if the observed value is larger than the upper bound of the credible interval.

The absolute error focuses on a particular point estimate, and the interval score focuses on an interval of a particular size. Both performance measures discard a great deal of information about the posterior distribution. The continuous ranked probability score takes account of much more information. To calculate the continuous ranked probability score, we start by obtaining the 1% quantile, 2% quantile, $\cdots$, 100% quantile, for each cell, using the approach discussed in Section 9.2. Each quantile is then used as a point estimate, and has an associated absolute error. The continuous ranked probability score is a weighted sum of these absolute errors.

Consider $\hat{Q}_{ast}(p)$ for $p = 0.01, 0.02, \cdots, 1$. The continuous ranked probability score is calculated as

$$\frac{2}{100} \sum_{p \in \{0.01, 0.02, \cdots, 1\}} \left\{ \left[I\left(\hat{Q}_{ast}(p) \le e_{ast}\right) p + I\left(\hat{Q}_{ast}(p) > e_{ast}\right)(1-p) \right] \left| \hat{Q}_{ast}(p) - e_{ast} \right| \right\}.$$

Here $\left| \hat{Q}_{ast}(p) - e_{ast} \right|$ is the absolute error of using the $100p\%$ quantile to forecast e_{ast}. Hence, if $\hat{Q}_{ast}(p) \le e_{ast}$, the weight for the absolute error is proportional to p, and if $\hat{Q}_{ast}(p) > e_{ast}$, the weight for the absolute error is proportional to $1-p$.

12.10   Results

We start with forecasts of life expectancy. Figure 12.10 shows forecasts of life expectancy at ages 0 and 65 for both sexes. Rather than our usual 50% and 95% intervals, we use 80% intervals, as is common in forecasting.

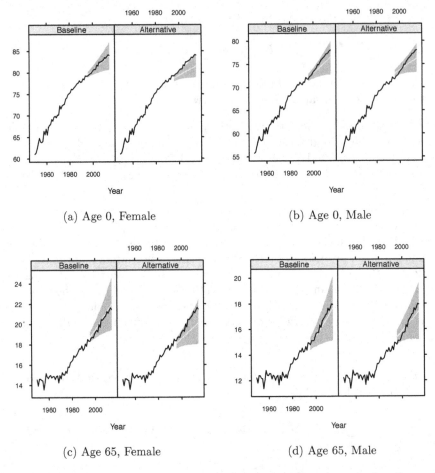

(a) Age 0, Female

(b) Age 0, Male

(c) Age 65, Female

(d) Age 65, Male

FIGURE 12.10: Forecasts of life expectancy for 1996–2015 from the baseline and alternative models. The white lines show the posterior medians, and the gray bands show 80% credible intervals.

The baseline model, which includes a sex-time interaction, clearly does better than the alternative model, which does not include a sex-time interaction. The posterior medians for the baseline model lie nearer to the observed values, and the credible intervals do not miss the observed values.

A particular problem with the forecast from the alternative model is that it can have big errors at the very start of the forecasting period. These sorts of early errors are known as jump-off errors. They occur when the statistical model does not accurately describe conditions in the years leading up to the forecast.

Jump-off errors are common in mortality forecasts, and indeed in demographic forecasts more generally. The best-known mortality forecasting method, the Lee-Carter method, includes special, rather ad hoc, post-estimation adjustments to reduce these errors. In our experience, however, jump-off errors can be reduced or eliminated by including enough interaction terms in a model, and hence improving model fit. The results in Figure 12.10 illustrate this.

Figure 12.11 presents performance measures for the two models. When calculating the interval score, we use 80% intervals. For all three performance measures, the baseline model wins in most cases.

12.11 Forecasting of Life Expectancy for 2016-2035

We next use all of the observed data for 1950-2015 to train the baseline model, and forecast life expectancy for the 20-year period of 2016-2035. Since we do not have exposure for the forecasting period, we do not convert the forecasted rates γ_{ast} into finite-population rates. We summarize the forecasted rates using life expectancies.

Figure 12.12 presents forecasts of life expectancy at ages 0 and 65 for both sexes. There is no sign of jump-off errors for the forecasts. In year 2015, the observed finite-population life expectancy is 84.0 years for females at age 0, 78.0 years for males at age 0, 21.5 years for females at age 65, and 17.9 years for males at age 65. The forecasts continue from these values, and suggest that life expectancy will continue to increase strongly for both sexes, particularly at the older ages.

12.12 Obtaining Forecasts of Life Expectancy*

To generate a single posterior draw of forecasts, we go through the following steps.

1. For the prior for time, plug the value of ω_δ and τ into Equations (12.14)– (12.16), and generate values for β^{time}.

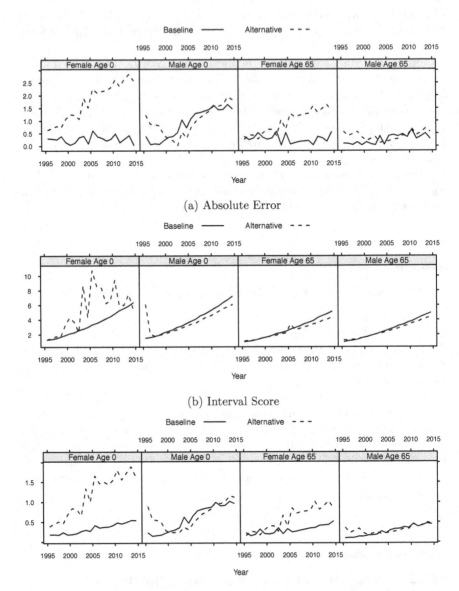

(a) Absolute Error

(b) Interval Score

(c) Continuous Ranked Probability Score

FIGURE 12.11: Performance measures of the baseline and alternative models using the heldback data.

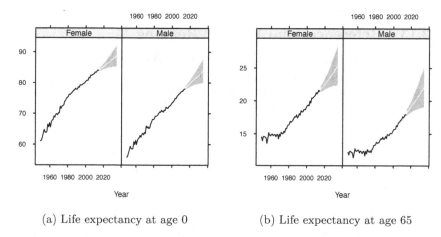

(a) Life expectancy at age 0   (b) Life expectancy at age 65

FIGURE 12.12: Forecasts of life expectancy for 2016–2035 from the baseline model. The white lines show the posterior medians, and the gray bands show 80% credible intervals.

2. For the prior for age-time interaction, plug values of ω_δ, τ, and ϕ into Equations (12.17)–(12.19), and generate values for $\beta^{\text{age:time}}$.

3. If sex-time interaction is included, generate values for $\beta^{\text{sex:time}}$ using the similar procedure of generating $\beta^{\text{age:time}}$.

4. Plug the values for β^{time}, $\beta^{\text{age:time}}$, and $\beta^{\text{sex:time}}$ if sex-time interaction is included, plus the values of β^0, β^{age}, β^{sex}, $\beta^{\text{age:sex}}$, and σ into Equation (12.9), and generate values for γ_{ast}.

5. To forecast finite-population quantities, do the following steps.

    - Generate death counts D_{ast} from Poisson distribution with means equal to exposure times the super-population mortality rates, $w_{ast}\gamma_{ast}$. Derive the finite-population mortality rates as $\gamma_{ast}^{\text{Fin}} = D_{ast}/w_{ast}$.

    - Summarize the finite-population mortality rates $\gamma_{ast}^{\text{Fin}}$ by life expectancy.

6. To forecast super-population life expectancy, summarize the super-population mortality rates γ_{ast} by life expectancy.

12.13   References and Further Reading

The death counts and exposures come from the *Human Mortality Database*, which is supported by the University of California, Berkeley, and the Max Planck Institute for Demographic Research. The data were downloaded from the website mortality.org on September 9, 2017.

The idea that computational problems may be a symptom of deeper conceptual problems with a model is referred to by Andrew Gelman as the "folk theorem of statistical computing" (andrewgelman.com/2009/05/24/handy_statistic). In our experience, computational problems can also be a symptom of programming errors.

Shmueli (2010) argues that different design principles are involved when building models that will be used for prediction and building models that will be used for explanation. She demonstrates how including extra variables in a model can lead to worse predictions, even when the variable is genuinely related to the outcome of interest.

Hyndman et al. (2008, Chapters 3 and 7) discuss the performance of damped local trend models, though they specify them in a slightly different way from our technique. Our prior for damping parameter ϕ is an example of a 'boundary-avoiding prior', as defined in Gelman et al. (2014, 313–318). The R package `forecast` is described in Hyndman and Khandakar (2008).

There is a very large literature on mortality forecasting. Booth and Tickle (2008) and Shang (2015) provide reviews; the latter paper also uses heldback data to compare the performance of some of the leading models. The best known method for mortality forecasting is that of Lee and Carter (1992). The revised Lee-Carter method described in Lee and Miller (2001) includes special adjustments to reduce jump-off error.

The models used by the United Nations to forecast world population have been subject to many evaluations using heldback data (Gerland et al., 2014).

Gneiting and Raftery (2007) describe measures for assessing the performances of forecasting models, including the absolute error, interval score and the continuous ranked probability score.

13

Health Expenditure in the Netherlands

The types of models that we have developed in Part III, with demographic arrays, hierarchical structures, age effects, time effects, and so forth, have applications outside the core demographic subjects of births, deaths, and migration. Examples include tax payments, energy use, violent crime, pension expenditures, hospital admissions, and diabetes prevalence.

Figure 13.1 illustrates why demographic models are more generally useful. The figure shows annual per capita health expenditure in the Netherlands, in 2011, for ages 0 to 95+. Average annual expenditure for a female aged 1–4 is €1,200. Average annual expenditure for a female aged 95+ is €46,400, or almost 40 times higher. If we want to explain or predict health expenditures, in the Netherlands or elsewhere, we need to pay attention to age. Moreover, it turns out that age is not the only demographic variable with substantial predictive power. Depending on the application, sex, education level, urban-rural residence, and other standard dimensions in demographic modeling also play a key role in explaining important social trends.

In this chapter, we use the example of health expenditure in the Netherlands to show how the framework of Part III can be applied to non-demographic events such as tax payments, crimes, and hospital admissions. We begin by setting out a conventional model for health expenditure in the Netherlands. We then show how a Bayesian version of the model addresses some of the limitations of the conventional model.

13.1 A Simple Expenditure Projection

Models of the relationship between demography and health expenditure typically take the form of a projection, that is, of a hypothetical statement on what would happen in the future, if various assumptions were met. Figure 13.2 shows a simple example.

At present the population contains 30 people, 10 of whom are old. Expenditure per person per year is $15 for young people, and $25 for old people. Total expenditure comes to $300 + $250 = $550. Per capita expenditure, averaging across young and old, is $550 ÷ 30 ≈ $18.

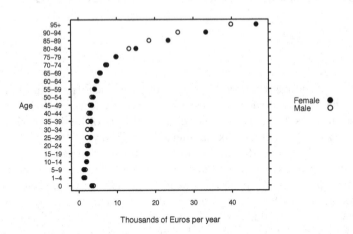

FIGURE 13.1: Health expenditure per capita in the Netherlands, in 2011, by age and sex.

	Now	Future
Young	20	20
Old	10	20

(a) Population

	Now	Future
Young	$15	$30
Old	$25	$50

(b) Expenditure per capita

	Now	Future
Young	$20 \times \$15 = \$300$	$20 \times \$30 = \$600$
Old	$10 \times \$25 = \$250$	$20 \times \$50 = \$1,000$

(c) Total expenditure

FIGURE 13.2: A simple health expenditure projection.

We assume—perhaps after doing a population forecast like the one described in Chapter 19—that the future population will contain 40 people, 20 of whom are old. These numbers imply population aging, in that the share of old people increases over time. We also assume that expenditure per person will double within each age group. Under these assumptions, total expenditure in the future is $600 + $1,000 = $1,600 and per capita expenditure is $1,600 \div 40 = $40.

Per capita expenditures within each age group increase by a factor of 2, while expenditures averaged across the whole population increase by a factor of $40 \div $18 \approx 2.2. The extra $2.2 - 2 = 0.2$ reflects population aging, or the shift of the population into the more expensive older age group. The number 0.2 is one possible measure of the contribution of demographic change to expenditure growth.

13.2   Expenditure Projections for the Netherlands

We move now to a more realistic example of a traditional expenditure projection, based on population and expenditure data for the Netherlands. Figures 13.3 and 13.4 show the data. For simplicity, we treat the estimated and projected population counts for 2003–2021 as known with certainty. (We describe in Section 13.5 how the model could be extended to incorporate uncertainty about population.) The population counts are plotted on a log scale, to highlight the changes occurring among older age groups. These changes are large in relative terms, but small in absolute terms, and would be obscured if plotted on an ordinary scale (see Section 12.1). The Netherlands population is set to change in much the same way as the hypothetical population in Figure 13.2. Younger age groups maintain their original size, while older age groups get larger.

Figure 13.4 shows health expenditures per capita, by age and sex, at four points in time, 2003, 2005, 2007, and 2011. The data are plotted on a log scale, which is why the differences across ages appear less dramatic than they do in Figure 13.1. Females and males generally have similar expenditures, except during the child-bearing years, and at the oldest ages. Expenditure per capita is rising within all age groups, though some groups, such as people aged 10–19 or 90+, have experienced faster growth than others.

A natural starting point for modeling future growth rates is to assume that they will be the same as historical growth rates. In many countries, calculating historical growth rates is difficult, because of a lack of data. However, the time series for the Netherlands, although tiny compared with most financial or economic time series, does have enough time points that allow us to calculate growth rates directly. Our baseline assumption is that expenditure within each age-sex group continues to grow at the average rate for 2003–2011. We obtain

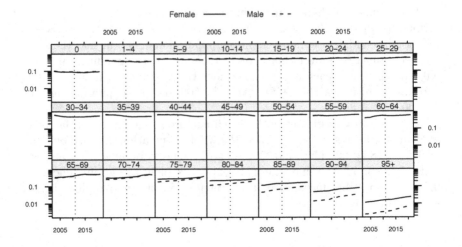

FIGURE 13.3: Estimated and projected population counts, in millions, for the Netherlands, by age, sex, and year. The values are plotted on a log scale. The vertical lines mark the year 2011, the last year for which we have health expenditure data.

average annual growth rates by fitting straight lines through logged per capita expenditures for each age-sex group—i.e. by fitting lines through the points in Figure 13.4. The median value is 6% per year, the minimum is 0.1%, and the maximum is 14%.

In addition to the baseline scenario, most traditional expenditure projections contain two or more alternative scenarios, with different assumptions about future growth rates. Alternative scenarios remind users that the future is uncertain, and serve as a type of experiment, showing what happens when one factor changes while everything else remains constant (Table 13.1). We include a "high" scenario, in which each age-sex group's future growth rate equals the corresponding historical growth rate plus 2 percentage points, and a "low" scenario, in which future growth rates equal historical rates minus 2 percentage points. We also include a "fixed" scenario, in which future growth rates are 0, and per capita expenditures are held at their 2011 level.

TABLE 13.1
Assumptions about future growth rates for per capita expenditures

Variant	Description
Baseline	Each age-sex group grows at its historical rate.
High	Historical rate plus 2%.
Low	Historical rate minus 2%.
Fixed	Zero.

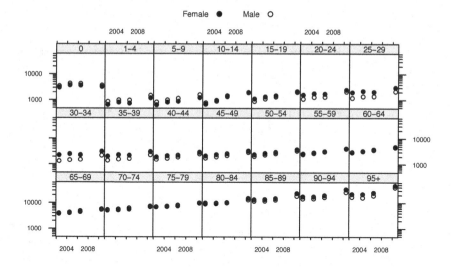

FIGURE 13.4: Health expenditure per capita in the Netherlands, by age, sex, and year, in 2003, 2005, 2007, and 2011. Expenditure is measured in 2011 Euros, and is shown on a log scale.

We generate future per capita expenditures for the period of 2012-2021 with the formula

$$e_t = e_{2011}(1+g)^{t-2011}, \tag{13.1}$$

where e_t is per capita expenditure for year t and g is the growth rate in per capita expenditure for the corresponding age-sex group. We then multiply future per capita expenditures by future population counts, in a larger version of the calculations in Section 13.1. From these numbers we calculate various summary measures. Expenditure projections are usually done using computer spreadsheets, but we did them with our R packages, which have functions for carrying out projections.

Results for the projections are shown in Figure 13.5. The first panel shows total health expenditure, summing up across all age-sex groups. Expenditure rises steadily under the baseline scenario, the high scenario, and even the low scenario. In contrast, under the fixed scenario, where non-demographic factors are held constant, so that the only source of change is population size and structure, the increases are small.

The results in the second panel of Figure 13.5 are derived by dividing the results in the first panel by population size. The second panel, in effect, removes the effects of population size. It turns out that growth rates without change in population size are close to those with change in population size.

The third panel shows the percentage of total expenditure that is spent on people aged 85+. By 2021, this group is projected to account for only 2.4% of the total population, but to account for roughly 15% of all health expenditure.

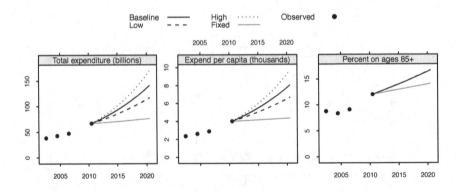

FIGURE 13.5: Traditional health expenditure projections. The first panel shows total health expenditure for the country. The second panel shows per capita expenditure, averaged across all age-sex groups. The third panel shows the percent of expenditure accounted for by people aged 85+. Results for the "baseline", "high", and "low" variants in the third panel are almost but not identical.

An odd feature of the third panel is that, in contrast to the previous two panels, the baseline, high, and low scenarios give results that are indistinguishable. The similarity is a mathematical consequence of the particular way in which the scenarios were set up. (Specifically, it follows from the fact that, when g and Δ are small, e.g. $g = 0.06$, $\Delta = 0.02$, or $\Delta = -0.02$, $(1 + g + \Delta)^t \approx (1 + g)^t (1 + \Delta)^t$. When taking ratios, the $(1 + \Delta)^t$ terms cancel.)

The phenomenon, apparent in Figure 13.5, of different scenarios giving dramatically different results according to some measures, but virtually identical results according to others, is common with expenditure projections. Sometimes this behavior reflects genuine features of the system being studied. But often it is an artefact of the way the scenarios are constructed. Scenarios that are designed to capture one sort of variation often fail to capture other sorts of variation.

A second problem with the scenario-based approach is that the meaning of the scenarios is unclear. How much more likely is the baseline scenario than the high scenario? How likely is it that actual expenditures will fall somewhere between the high and low scenarios?

Both these problems can be addressed by taking a more Bayesian approach to expenditure projections. To do this, we must first build a statistical model of per capita expenditures.

13.3 A Statistical Model for Per Capita Expenditures

Let γ_{ast} be expenditure per capita for age a, sex s, and year t. When economists and others model expenditures, or other non-negative monetary variables such as wages or income, they typically work on a log scale, which helps pull in high values, and generally leads to a better-behaved dataset. A natural model for per capita expenditures is therefore something like

$$\log \gamma_{ast} \sim \mathrm{N}(\beta_0 + \beta_a^{\mathrm{age}} + \beta_s^{\mathrm{sex}} + \beta_t^{\mathrm{time}} + \beta_{as}^{\mathrm{age:sex}} + \beta_{at}^{\mathrm{age:time}}, \sigma^2). \qquad (13.2)$$

Equation (13.2) looks different from the models in Chapters 11 and 12 in that the β appear in the first level of the model, rather than the second. If, however, we use y_{ast} to denote total expenditures in cell ast, and use w_{ast} to denote exposure, so that $\gamma_{ast} = y_{ast}/w_{ast}$, then the model of Equation (13.2) is mathematically equivalent to

$$y_{ast} = \gamma_{ast} w_{ast} \qquad (13.3)$$
$$\log \gamma_{ast} \sim \mathrm{N}(\beta_0 + \beta_a^{\mathrm{age}} + \beta_s^{\mathrm{sex}} + \beta_t^{\mathrm{time}} + \beta_{as}^{\mathrm{age:sex}} + \beta_{at}^{\mathrm{age:time}}, \sigma^2). \qquad (13.4)$$

The model of Equations (13.3) and (13.4) looks more like the models from Chapters 11 and 12. The most important difference is that y_{ast} is exactly predicted by γ_{ast} and w_{ast}, rather than being drawn from a distribution. Any unpredictability in the y_{ast} comes from the prior model described in Equation (13.4). Our reason for taking this approach is purely pragmatic. With the data at hand, there is no way of distinguishing between errors due to limitations of the prior model and errors due to random variation in y_{ast}.

To decide which main effects and interactions to include in the prior model, we carried out a decomposition similar to the one described in Sections 12.4 and 12.5. We do not show the results here, but we found that age effects, sex effects, time effects, age-sex interactions, and age-time interactions were all large enough to warrant including them in the model, but sex-time interactions were not.

The age-sex profile for Dutch health expenditure looks remarkably like the typical age-sex profile for mortality (e.g. Figure 12.1). We therefore use the same set of priors for age effects, sex effects, and age-sex interactions that we used for mortality in Chapter 12: a local trend model with a covariate indicating infants, a normal model, and a local level model.

The most important, and most difficult, part of specifying the prior model for per capita health expenditure is choosing appropriate priors for the time effect and age-time interaction. Ideally, we would like to use something like a local trend model. The local trend model, however, has far too many parameters to be estimated from a time series with only four points. Informative priors would help, but specifying appropriate priors is challenging.

Instead, we use a more elaborate version of the straight-line model from the traditional projection. We assume that the time effect and age-time interactions have the form

$$\beta_t^{\text{time}} = \alpha^{\text{time}} + \delta^{\text{time}} t + e_t^{\text{time}} \tag{13.5}$$

$$\beta_{at}^{\text{age:time}} = \alpha_a^{\text{age:time}} + \delta_a^{\text{age:time}} t + e_{at}^{\text{age:time}} \tag{13.6}$$

$$e_t^{\text{time}} \sim \text{N}(0, \omega_{\text{time}}^2) \tag{13.7}$$

$$e_{at}^{\text{age:time}} \sim \text{N}(0, \omega_{\text{age:time}}^2). \tag{13.8}$$

Time effect β_t^{time} follows a straight line with intercept α^{time} and slope δ^{time}, but with random deviations upwards or downwards described by the errors e_t^{time}. Similarly, each age-time effect $\beta_{at}^{\text{age:time}}$ follows its own straight line, with random fluctuations upwards and downwards.

We use weakly informative priors for all the parameters in Equations (13.5)–(13.8). The parameters in Equations (13.5)–(13.8) are estimated from the expenditure data together with the rest of the parameters in the model. We obtain a posterior distribution for the parameters, based on historical data.

13.4   Model Checking via Replicate Data

Our statistical model has some flexibility, but the assumption of straight lines might still be too strong. Before proceeding to the projection, we check whether the model adequately captures the relevant features of the historical data. We do this by using the model to generate replicate data, and comparing the replicate data to the actual data. (For other discussions of replicate data, see Sections 9.6.3 and 11.5.2.) If the actual and replicate data look similar, then we assume that the model is performing well enough to use.

We carry out two replicate data checks. The first focuses on growth rates in health expenditure, and the second on the percent of expenditure accounted for by people aged 85+.

Both checks use the same replicate data. After fitting the model, we randomly select from the posterior sample 20 sets of β and σ. We plug each set of β and σ into Equation (13.4), to generate per capita expenditures.

To conduct the first test, we calculate growth rates in total health expenditure using actual and replicate data. We combine actual population data with actual per capita expenditure data to calculate actual total expenditure in 2003, 2005, 2007, and 2011, and then calculate average expenditure growth for the periods 2003–2005, 2005–2007, and 2007–2011. We then repeat the same process 20 times, each time replacing actual per capita expenditure data with one set of replicate per capita expenditure data. After carrying out these calculations, we have three actual growth rates and $20 \times 3 = 60$ replicate growth rates. The actual and replicate growth rates are plotted in Figure 13.6.

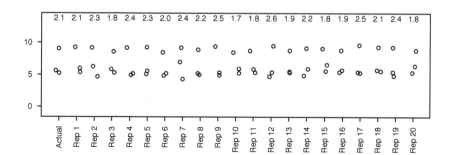

FIGURE 13.6: Growth rates in total health expenditure calculated from actual and replicate data on per capita expenditure. The growth rates are calculated for the periods 2003–2005, 2005–2007, and 2007–2011. The values along the top of the graph are standard deviations for the three growth rates.

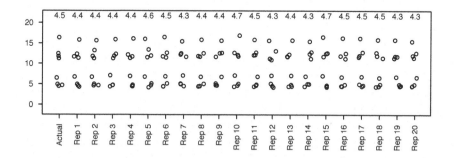

FIGURE 13.7: The percent of total health expenditure going to people aged 85+, calculated from actual and replicate data. Each dot represents one combination of sex and year. The values along the top of the graphs are standard deviations.

The actual growth rates look like they could have been drawn from the same distribution as the replicate growth rates. This suggests that the model is doing a reasonable job of describing historical variability in overall expenditure. We hope that if the model does a reasonable job in describing historical variability, it will do a reasonable job of describing future variability.

For our second test, we calculate the percentage of overall expenditure going towards people aged 85+ in 2003, 2005, 2007, and 2011, using the actual data and the 20 sets of replicate data. We do the calculations separately for each combination of sex and year, to yield $2 \times 4 = 8$ percentages from each dataset. These percentages are plotted in Figure 13.7. Once again, the results based on actual data are indistinguishable from the results based on replicate data. This increases our confidence in the forecasted results on percentages accounted for by people aged 85+.

13.5   Revised Expenditure Projections

After fitting the statistical model to the historical data, we forecast future per capita expenditures. The forecasts come in the form of a sample from the posterior distribution. To construct the expenditure projections, we process the sample in the same way that we processed the four scenarios in the traditional projection. We take each set of forecasted per capita expenditures, multiply it by future population counts to obtain future expenditures, and then calculate values for overall expenditure, per capita expenditure, and expenditure on ages 85+. We thus obtain a sample from the posterior distributions of overall expenditure, per capita expenditure, and expenditure on ages 85+. (Section 9.3.1 discusses obtaining a posterior distribution for derived quantities.)

Extending the calculations to include uncertainty about future population counts would easy, provided we had a probabilistic population forecast available to us. Rather than using the same set of population counts with each set of forecasted per capita expenditures, we would simply draw a set of population counts from the posterior sample and use that. Each randomly-generated set of per capita expenditure values would be combined with a different randomly-generated set of population values.

Figure 13.8 summarizes the three posterior distributions obtained from our revised expenditure projections. The summaries use 80% intervals rather than 95% intervals: 80% intervals are commonly used in economic forecasting.

As well as forecasting future expenditure values, the model can be used to estimate values for years in the past without data. In fact, as we saw in Section 9.5, Bayesians do not make a sharp distinction between estimates and forecasts. Figure 13.8 shows historical estimates along with the forecasts.

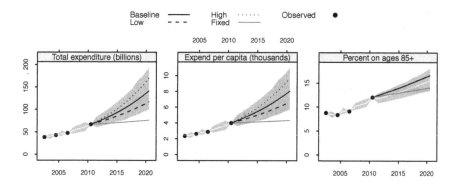

FIGURE 13.8: Health expenditure projections based on the statistical model. The gray bands show 80% credible intervals from the statistical model, and the white lines represent posterior medians. For comparison, results from the scenario-based approaches are repeated from Figure 13.5.

In the first two panels of Figure 13.8, the posterior medians from the Bayesian model sit on top of the "baseline" scenario from the traditional projection. In the third panel, however, the posterior medians differ from the "baseline" scenario. It seems that using a more complicated model of correlations among age-sex groups, as the Bayesian model does, leads to different point estimates of the percentage of expenditure going to people aged 85+.

If we expand our assessment to include uncertainty, then the contrast between the Bayesian and traditional approaches is stronger. The Bayesian model provides probabilistic statements about what would happen if historical trends, and historical levels of variability, were to continue into the future. The empirical status of the "high" and "low" scenarios is less clear. It is tempting to interpret them as approximate credible intervals. In the first two panels, they behave like 65% credible intervals: according to the Bayesian model, there is an approximately 65% chance that total expenditure and per capita expenditure fall within the projected values under the "high" and "low" scenarios. The analogy breaks down, however, when we turn to the third panel. Here the "high" and "low" scenarios contain only a tiny fraction of possible outcomes. The Bayesian model, in contrast, continues to give sensible results.

The information that the Bayesian projections provide about uncertainty has policy relevance. When the outcome is virtually certain under current policies, the rational response is usually to act now. When the outcome is merely likely, the rational response may be to defer a decision until more information is available.

Bayesian projections can be used as inputs to a formal decision-making process. Several possible policy options are identified. A "loss function" is specified, describing the losses that would be incurred under every possible combination of policy option and future health expenditure. For each policy option, the average loss across all posterior draws of projected expenditures is calculated. The best policy option is the one with the smallest average loss.

13.6   Forecasting Policy Outcomes

In the 1980s and 1990s, demographers and economists in rich countries presented governments with alarming projections about future expenditures on pensions and health care. Many governments responded by changing the policies surrounding pensions and health care, to try to constrain future expenditures. The government responses were an example of feedback from projections to the phenomena being projected. Scientists constructing projections of greenhouse gas emissions or diabetes prevalence typically hope for the same sort of feedback.

The possibility for policy feedbacks complicates the interpretation of expenditure projections, whether traditional or Bayesian. There is always some sort of implicit "if historical policy settings were maintained" condition attached to the projections. This is an unavoidable consequence of the fact that humans understand forecasts and react to them.

To some extent, the same is true of demographic forecasts of mortality, fertility, migration, and population. If policy makers, or people more generally, do not like the forecasted values, they can take action to change them. However, altering demographic trends is difficult. Migration is the demographic series most amenable to deliberate control, but even here, policies often have less effect on overall numbers than their proponents hope for. While governments may have some control over the number of people entering the country, for instance, they have less control over the number leaving.

13.7   References and Further Reading

The data on health expenditure come from the table *Current Health Spending by Age* in the OECD database *OECD.stat*, and were downloaded on November 24, 2017. The original expenditure data were in current prices. We adjusted them to 2011 prices using the Dutch consumer price index from the table *Key Short-Term Economic Indicators: Consumer Prices - Annual inflation* of the OECD database *OECD.Stat*, downloaded on June 3, 2016. The population estimates and projections come from the tables *Population on 1 January (tps00001)* and *Population projections (tps00002)* on the Eurostat database. The data were downloaded on November 26, 2017.

Cutler and Sheiner (1998) present an early, influential discussion of the relationship between demographic change and health expenditure. The authors argue that, despite widespread worries about the effects of population aging, demographic change is not the most important determinant of expenditure growth. Dieleman et al. (2017) decompose increases in health care expenditure in the United States into five factors, two of which are population growth, and population aging. Lee and Miller (2002) present probabilistic forecasts of health expenditures in the United States.

Many governments produce regular reports on the long-term fiscal health of the country that involve numerous expenditure projections. Australia's *Intergenerational Report* available from the Australian Treasury's website, is a representative example. Kronenberg (2009) use the relationship between age and energy use to project energy use and greenhouse gas emissions in Germany. Wild et al. (2004) project global diabetes prevalence.

Part IV

Inferring Demographic Arrays from Unreliable Data

14

Inferring Demographic Arrays from Unreliable Data

In Part III, we treated demographic arrays of counts or totals as directly observed, with no measurement error. We focused on inferring the underlying rates, probabilities, and means, along with the prior model.

In Part IV, we no longer assume the absence of measurement error. Instead, we infer the true birth counts, death counts, or other demographic quantities along with all the other unknowns. We use one or more datasets, allowing for the possibility that these datasets are incomplete or unreliable. We deal with unreliable data by specifying explicit models for measurement error.

In this chapter, we give an overview of the expanded framework. In the next two chapters, we illustrate the framework with case studies from Iceland and Cambodia.

14.1 Summary of the Framework

The framework of Part IV is summarized in Figure 14.1. The components are:

Demographic array Y. True counts or totals, organized into a demographic array. The Y of Part IV is identical to the Y of Part III except that it is not observed directly.

Exposure array W. Data on exposures. Identical to the W in Part IV. As with Part III, not all models include W.

Array γ of rates, probabilities or means. Identical to γ in Part III.

Vector ϕ of parameters from priors. Identical to ϕ in Part III.

Datasets $X_1, \cdots, X_M$. Observed arrays of counts or totals that are used to estimate the true counts or totals. The datasets may lack dimensions or categories that are present in Y.

Vectors $\Omega_1, \cdots, \Omega_M$ of parameters from data models. Each dataset has an associated data model describing how the data are generated from the true counts or totals.

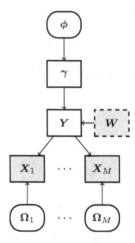

FIGURE 14.1: Inferring an array of counts or totals, and associated rates, probabilities, or means, from one or more datasets. Y is the true array of counts or totals; γ is an array of rates, probabilities, or means; ϕ holds parameters from the priors for γ; and W is an optional array of exposures. The $X_1, X_2, \cdots, X_M$ are datasets, and $\Omega_1, \Omega_2, \cdots, \Omega_M$ are parameters from the associated data models. The only elements of the framework that are observed are $X_1, \cdots, X_M$ and W.

Allowing for one or more unreliable datasets gives us much more flexibility than requiring a single perfect dataset. In return for this extra flexibility, we need to specify explicit data models. In other words, for each dataset X_m, we need to set out a model describing the probability of observing the dataset given the truth Y. If, for instance, we believed that each person in the population had a 96% chance of being captured by the census, regardless of age, sex, or any other attribute, then our model might be

$$x_i^{\text{cens}} \sim \text{binomial}(y_i, 0.96) \tag{14.1}$$

where y_i is a cell within Y and x_i^{cens} is the corresponding cell within the census dataset.

Using data models has a big practical advantage: it allows us to use datasets that are less detailed than Y. A dataset X_m does not need to include all the dimensions that are present in Y. Table 14.1 shows some examples. As can be seen in the third row of Table 14.1, X_m is not allowed to have extra dimensions that are not present in Y. This never causes any practical problems, however, since any extra dimensions in X_m can always be collapsed (Section 4.10).

Similarly, X_m can omit categories that appear in Y, or can use coarser categories, but cannot have extra categories or finer categories. Table 14.2 shows some examples.

If Y has dimensions that X_m does not, then, as part of the estimation process, we can collapse the extra dimensions in Y before supplying Y to the

TABLE 14.1
Combinations of dimensions in dataset X_m and array Y

Dimensions in X_m	Dimensions in Y	Status
age, sex, time	age, sex, time	Permitted
age, time	age, sex, time	Permitted
age, sex, time, region	age, sex, time	Not permitted

data model for X_m. Similarly, if a dimension of Y has categories that the corresponding dimension of X_m does not, we can drop the extra categories. If a dimension of Y uses finer categories than the corresponding dimension of X_m, we can collapse the categories to match the categories in X_m.

TABLE 14.2
Combinations of categories in dataset X_m and array Y

Dimension	Categories in X_m	Categories in Y	Status
region	A, B, C	A, B, C	Permitted
region	A, B	A, B, C	Permitted
region	A, B, C, D	A, B, C	Not permitted
age	0–4, 5–9	0–4, 5–9	Permitted
age	0–4	0–4, 5–9	Permitted
age	0–9	0–4, 5–9	Permitted
age	0–4, 5–9, 10–14	0–4, 5–9	Not permitted
age	0, 1–4, 5–9	0–4, 5–9	Not permitted

Traditional demographic models have much less tolerance for inconsistent classifications. Instead, analysts have to reformat datasets and impute values until the data have the same structure as the quantities being estimated. The process of tidying the input data is slow and error-prone. Some datasets may be too incomplete to fit the required structure, and must be left out of the estimation process, or incorporated in some special hand-crafted way. The imputation and extrapolation processes also typically involve treating imputed values as if they were observed, which, as discussed in Section 9.4, leads to estimates that are spuriously precise.

The framework of Part IV makes no strong distinctions between estimation and forecasting. A forecast is just an example of Y having extra categories— in this case time points or time periods that are not present in the input data.

The models of Part IV produce the same super-population quantities describing Y, that is, the same γ and ϕ, as the models of Part III. But, in addition, they also produce output from the data models. Output from the data models can include probabilities that people or events will be captured by a data source, rates of over-reporting, and estimates of whether coverage is improving or deteriorating over time.

The other big difference between the output from the models of Part IV and the models of Part III is that the models of Part IV produce estimates of Y. These estimates are finite-population quantities. They try to capture the actual number of events that occurred, or the actual number of people present, rather than underlying propensities.

The estimates of finite-population Y, just like the estimates of super-population γ, take the form of a sample from the posterior distribution. We can therefore calculate virtually any uncertainty measure for Y that we like.

The estimates of Y, the estimates of γ, and the estimates of all the other unknown quantities are also mutually consistent. Our estimates of the number of deaths, for instance, are formed jointly with our estimates of the death rates. Moreover, the level of uncertainty we have about the number of deaths reflects the level of uncertainty we have about death rates, and vice versa.

Example 14.1. Figure 14.2 illustrates a hypothetical example in which true death counts need to be estimated from registered deaths and hospital data. This example is modified from Example 10.1 of Part III. The Poisson model for deaths and the prior for γ are the same as those of Example 10.1, except that Figure 14.2 uses dimensions "sex" and "year" rather than "sex" and "region". The new feature of Figure 14.2 is the multiple datasets and associated data models.

The first dataset is registered deaths. The registration process is imperfect, with registered deaths usually understating, but sometimes overstating, the true number of deaths. The relationship between registered deaths and true deaths is modeled using

$$x_{st}^{\text{reg}} \sim \text{Poisson}(\lambda y_{st}). \tag{14.2}$$

The λ in Equation (14.2) measures the expected number of registered deaths for each actual death.

The second dataset is hospital records. Hospital records do not include information on the sex of the deceased. Because some deaths occur outside hospitals, hospital records typically understate the true number of deaths. The relationship between hospital records and actual deaths is modeled using

$$x_t^{\text{hos}} \sim \text{binomial}(y_{Ft} + y_{Mt}, \pi). \tag{14.3}$$

The π in Equation 14.3 measures the probability that an actual death will be captured by hospital records. Since the hospital data do not distinguish between sexes, deaths for females and males from Y are summed before they are supplied to the data model.

The death counts, exposure, and death rates extend through to the year 2020, while the data only go to 2015, so the model combines estimation and forecasting.

			year		
			2010	2015	2020
sex	Female		?	?	?
	Male		?	?	?

Parameters ϕ (with μ: ? and σ: ?) — Rates γ

		year		
		2010	2015	2020
sex	Female	?	?	?
	Male	?	?	?

Counts Y

		year		
		2010	2015	2020
sex	Female	18.3	15.2	16.7
	Male	33.8	19.1	25.2

Exposure W

		year	
		2010	2015
sex	Female	3	9
	Male	8	7

Registered deaths X_1

year	
2010	2015
10	14

Hospital records X_2

FIGURE 14.2: A simple example of a model using the framework of Part IV. □

14.2 Data Models

Data models attempt to describe the process by which the data are generated from the true counts or totals. For instance, given the true number of people who are unemployed, and their distribution across different parts of the country, a data model might predict how many people register for unemployment benefits, and how the registrations are distributed across the country. The model needs to allow for systematic biases, such as any longstanding variation in registration rates across regions, but also random variation, such as idiosyncratic rises or falls in registration rates.

Constructing a data model is easiest when some highly reliable data source is available to calibrate the model against. For instance, it is easy to identify systematic biases in registration rates for unemployment benefits if accurate measures of unemployment are available. By comparing the number of people who register for benefits in each region with the number of people who are in fact unemployed, we can calculate region-specific registration rates. The usual source of accurate measures of unemployment is household labor force surveys.

A highly reliable source is not always available, however. An important example is information on numbers of migrants. In countries without accurate population registers (i.e. most of the world outside northern Europe) the best source of information on people who moved within the country or from other countries is the population census. But population censuses are always at least five years apart, and if migration rates are changing quickly, data from the most recent census may no longer be relevant. Moreover, as discussed in Section 6.1, population censuses in many countries have an uncertain future.

In the absence of a gold standard, we can follow an approach that economic statisticians refer to as "data confrontation". We compare multiple datasets that we know to be imperfect in the hope that the strengths of some datasets can be used to compensate for the weaknesses of others. For instance, based on descriptions of how the data are collected, we might believe that one dataset does a good job of capturing overall numbers, but measures region inaccurately, while another dataset only captures certain groups, but measures region accurately.

Data models can be used to encode beliefs about the strengths and weaknesses of datasets. For instance, because different age groups have different propensities to pay taxes, we might assume that the coverage of the tax system varied by age. When building a data model for a tax dataset, we would use a prior for the age dimension that permitted large differences between age groups. However, if tax rules and their implementation had not changed for several years, we might assume that, within each age group, coverage rates had remained more or less the same over time. In this case, we would use a prior for the time dimension that permitted only small differences between years.

Our framework—and the R software implementing it—are much less prescriptive about data models than they are about models for Y. The only restriction placed on the models is that they must predict the data given Y. In general, however, data models need to be relatively simple. If they have too many unknowns, and provide too much flexibility, we may be unable to distinguish between genuine features of Y and artefacts of the measurement process.

14.3 Applications

The models of Part IV can be used in any application where the models of Part III would have been used if a comprehensive and reliable dataset had been available. If, for instance, we wanted to estimate and forecast fertility rates, but instead of one set of good quality birth counts, we had one or more incomplete or unreliable sets, then we would use the models of Part IV.

Sometimes we are interested in Y itself, rather than the parameters governing Y. For instance, we might want to estimate the number of migrations, the number of people with diabetes, or the number of people with tertiary qualifications from several datasets.

In some cases, it is not Y or the parameters governing Y that we are most interested in, but rather the data sources and their relation to Y. We might, for instance, be trying to assess the coverage of birth registrations. The models of Part IV provide us with finite-population estimates of the proportion of actual births that were registered in each year.

14.4 References and Further Reading

The models of Part IV overlap with measurement error models in statistics. A Bayesian book on measurement error models is Gustafson (2003), and a non-Bayesian one is Buonaccorsi (2010).

Raymer et al. (2013) is a pioneering study in which expert judgement is used to construct data models for European data on international migration.

15

Internal Migration in Iceland

In our first application with noisy data we examine the special but important case where the noise has been added to the data deliberately. The reason for adding the noise is to protect confidentiality.

Consider a dataset containing the record of a 32-year-old unmarried male migrating from small region A to small region B. A data user who was familiar with small region A or B might recognize the 32-year-old unmarried male in the data, and hence find out other, less public, information about the person, such as income or health status. Many people, including possibly the 32-year-old male, would consider this a breach of privacy.

To avoid such breaches, statistical agencies often do not release raw data to the public, but instead release confidentialized data in which the true values have been partly obscured. For instance, rather than report that the unmarried male was age 32, for instance, the confidentialized data might only record that he was aged 30–34.

Confidentialized data can be analyzed within the framework of Part IV. The *un*confidentialized data, which we assume the analyst never sees, is the array of true values Y. The confidentialized data, which the analyst does see, is the dataset X. The aim of the analysis is to infer the rates or probabilities underlying Y. To do so, the analyst also needs to infer Y.

Confidentialized data are a special case of analysis of noisy data because the process generating the data is known. Building a data model for a known process is easy. In future chapters, we will turn to the more difficult general case where the process generating the data is not known, so that the data model is imperfect and can have parameters that need to be estimated.

The particular set of confidentialized dataset that we examine in this chapter is data on migration between regions of Iceland. We begin by describing the data and the confidentialization process. We then construct a model for the underlying migration rates, and jointly estimate migration counts and rates. We compare the results from our main analysis with an alternative analysis, in which we we treat the confidentialized data as noise-free.

15.1 Internal Migration in Iceland

Icelandic official statistics divide the country into eight regions. In this chapter, we model migration between these eight regions, for the period 2006–2015. We model migration by one-year age group, by sex, by origin, by destination, and by time. Migration is measured using movements arranged in an origin-destination format, as described in Section 4.3. Movements within regions are ignored, so flows where the region of origin equals the region of destination are by definition zero.

Cell counts in the Icelandic migration data are small. The population of Iceland in 2015 was 329,000. Across all 10 years of the (confidentialized) migration data, excluding cells where the origin equals the destination, the average migration cell count is 0.97, the maximum cell count is 33, and 79% of cells are equal to zero. Almost two thirds of Iceland's population lives in the Capital Region, and flows to and from that region are over ten times larger, on average, than flows between the other regions. The average cell size for flows between the smaller regions is 0.25.

Figure 15.1 shows direct estimates of migration rates between four selected regions. Separate estimates are shown for each age group, averaging over both sexes and the entire period 2006–2015. Despite all this averaging, the estimates are subject to substantial random variation. Flows to and from the Capital Region are large enough for an age profile to be visible, with peaks at the youngest ages and around age 25. Flows between other regions are so erratic that patterns are difficult to discern.

15.2 Confidentialization by Random Rounding to Base Three

Statistical agencies often confidentialize data using a technique known as "random rounding to base three". Under this rule, a value n is rounded as follows:

- If n is already divisible by 3, leave it unchanged.

- If n is not divisible by 3, find the two nearest values that *are* divisible by three. With probability 2/3 choose the closer of these two values, and with probability 1/3, choose the other value.

Figure 15.2 illustrates random rounding for numbers 0 to 6. If, for instance, the unconfidentialized value is 1, then there is a 2/3 chance that the confidentialized value will be 0, and a 1/3 chance that it will be 3. If the unconfidentialized value is 0, then the confidentialized value will always be 0.

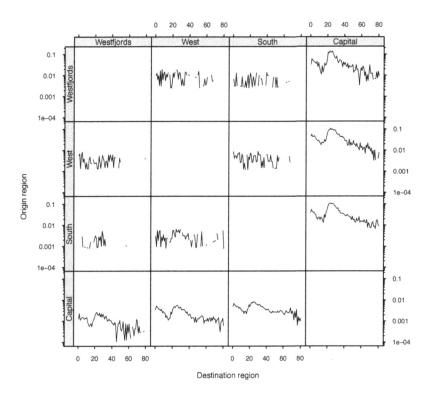

FIGURE 15.1: Direct estimates of migration rates between four selected regions of Iceland, for the period 2006–2015. Each row shows a different origin, and each column shows a different destination, so that, for instance, the bottom left panel shows migration from Capital Region to Westfjords Region. The rates are calculated from confidentialized data, averaged over females and males, and averaged over the entire period. Values are plotted on a log scale, and rates of 0 are omitted.

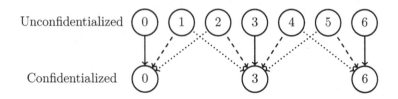

FIGURE 15.2: Random rounding the numbers 0 to 6 to base three. Solid lines denote probability 1, dashed lines denote probability 2/3, and dotted lines denote probability 1/3.

Statistics Iceland in fact publishes its migration data unrounded. To obtain confidentialized data for this chapter, we rounded the values ourselves. This allows us to see what data look like before and after rounding.

Figure 15.3a shows the age profile for migration from South Region to Capital Region by females in 2015, before and after rounding. Confidentialization produces a few distortions, particularly at the lowest and highest ages, where the profile takes on a step-like pattern. Overall, however, the shape of the distribution is left more-or-less intact.

Figure 15.3b shows the corresponding age profile for migration from South Region to West Region. The numbers in this case are much smaller, with many zeros. Here, confidentialization has a dramatic effect. Random rounding seems to have stripped almost all the information out of the data.

15.3   Overview of Model

Our overall model has five components, summarized in Figure 15.4. There is a system model describing patterns in the true, unconfidentialized migration counts. The system model includes exposure, which we treat as error-free. The confidentialized counts are jointly determined by the true counts and the data model. Unlike in most noisy-data applications, we know the correct data model, so we regard the data model as observed.

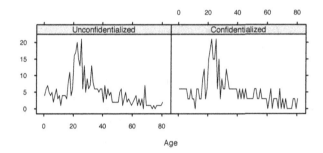

(a) Migration from South Region to Capital Region.

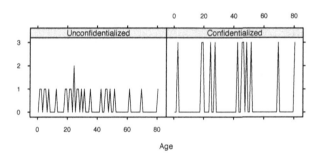

(b) Migration from South Region to West Region.

FIGURE 15.3: Data on internal migration from South Region, for females in 2015—with and without confidentialization by random rounding to base 3. The graph for migration to Capital Region uses a different vertical scale from the graph for migration to West Region.

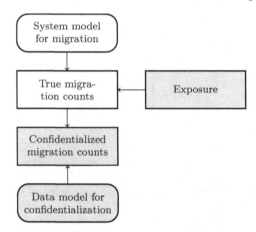

FIGURE 15.4: An overview of our model for internal migration in Iceland. Straight-edged rectangles represent demographic arrays, and rounded rectangles represent models. Gray shapes are observed; everything else, including the true, unconfidentialized migration counts, must be inferred.

15.4 System Model

Let y_{asijt} be the true, unobserved number of movements between regions i and j for age a, sex s, and year t. We assume that y_{asijt} has distribution

$$y_{asijt} \sim \begin{cases} \text{Poisson}(\gamma_{asijt} w_{asit}) & \text{if } i \neq j \\ 0 & \text{if } i = j, \end{cases} \tag{15.1}$$

where w_{asit} is exposure. The fact that the true migration counts are unobserved is not recognized in any way by the system model. The job of the system model is solely to describe demographic behavior. It says nothing about measurement processes.

The 0 in Equation (15.1) is a structural zero (see Section 4.3 for an introduction to structural zeros). A structural zero is not really part of the dataset, but is instead a logical consequence of the way the problem is set up. In Equation (15.1) it is a consequence of the convention that movements within a region do not count as internal migration.

The indices for the exposure term w_{asit} include the origin region i but not the destination region j. It is the population of region i, not region j, that is exposed to the risk of migrating from region i to region j.

We model migration rates γ_{asijt} using

$$\log \gamma_{asijt} \sim \text{N}(\beta^0 + \beta_a^{\text{age}} + \beta_s^{\text{sex}} + \beta_i^{\text{orig}} + \beta_j^{\text{dest}} + \beta_t^{\text{time}} + \beta_{as}^{\text{age:sex}} + \beta_{ij}^{\text{orig:dest}}, \sigma^2), \tag{15.2}$$

$i \neq j$. The model includes age, sex, origin, destination, and time main effects, an interaction between age and sex, and an interaction between origin and destination. The model is only defined for cells $asijt$ without structural zeros.

The origin effect captures differences in regions' propensity to send migrants, and the destination effect captures differences in regions' propensity to receive migrants. The origin-destination interaction effect captures the residual differences in rates of migrations between regions, after accounting for the origin and destination main effects.

We chose the specification described in Equation (15.2) after decomposing and graphing the confidentialized data. The process was essentially the same as the one we describe for the Portuguese data in Sections 12.4 and 12.5.

We use a local trend model (Section 11.3.4) for the age effect, a local level model (Section 8.5.3) for the age-sex interaction, and a local level model for the time effect. We use a local level model, rather than a local trend model, for time because there is no clear upward or downward trend in overall migration rates. Standard deviation terms for the age, age-sex, and time priors all have the same half-t weakly informative priors, as does σ. The intercept has a $N(0, 10^2)$ prior, and the sex term has a $N(0, 1)$ prior.

The origin effect has prior

$$\beta_i^{\text{orig}} \sim N(0, \tau_{\text{orig}}^2) \tag{15.3}$$

where τ_{orig} has our standard half-t weakly informative prior. The destination effect and the origin-destination interaction have similar priors. The origin-destination interaction $\beta_{ij}^{\text{orig:dest}}$ includes ij where origin i differs from destination j.

15.5 Data Model

Let x_{asijt} be the confidentialized version of the true, unconfidentialized count y_{asijt}. If y_{asijt} is 2, for instance, then x_{asijt} would be 0 or 3. We need a data model showing how x_{asijt} is derived from y_{asijt}. To do this, we need to write out the rules for random rounding described in Section 15.2 in the form of a likelihood. One way of doing so is

$$p(x_{asijt}|y_{asijt}) = \begin{cases} 1 & \text{if } x_{asijt} = y_{asijt} \\ 2/3 & \text{if } |x_{asijt} - y_{asijt}| = 1 \\ 1/3 & \text{if } |x_{asijt} - y_{asijt}| = 2 \\ 0 & \text{otherwise.} \end{cases} \tag{15.4}$$

(The paired vertical lines denote "absolute value", so that, for instance, $|3 - 5| = 2$.) The x_{asijt} in Equation (15.5) are all divisible by 3, since they have all been through the rounding process. The y_{asijt}, in contrast, can be any whole number.

The data model is probabilistic in that values for x_{asijt} are generated randomly. The model does not, however, contain any unknown parameters. There is nothing in the model that needs to be estimated after we have seen the data.

15.6   Estimation

We fit the model using our R packages. The estimation function starts with some rough guesses at the unconfidentialized counts and the parameters of the system model. It then alternates between updating its estimates of the unconfidentialized counts, conditional on the parameters of the system model, and updating its estimates of the parameters of the system model, conditional on the unconfidentialized counts. The process continues until convergence.

At the end of the estimation process, we have a set of draws from the joint distribution of the migration counts, migration rates, and other parameters. The term 'joint' is important. Because the unconfidentialized counts and the system model parameters are calculated altogether, they are all consistent with each other. If migration counts for a particular region are particularly uncertain, for instance, then the migration rates for that region will also be uncertain.

15.7   Results for Unconfidentialized Migration Counts

We begin with results for the unconfidentialized migration counts. Displaying results for the unconfidentialized counts is tricky. The counts are small and are closely bunched around the confidentialized values. Our standard summary measures, credible intervals and posterior medians, do not work well with these sorts of distributions. We therefore take an alternative approach, shown in Figure 15.5.

Figure 15.5 shows posterior distributions for the unconfidentialized counts. The counts are for females aged 20, migrating out of South Region, for each year between 2006 and 2015. The upper graph shows migration to Capital Region, and the lower graph shows migration to West Region. Confidentialized values are marked with an ×. The results in the top left panel, for instance, indicate that there is an approximately 0.1 probability that the unconfidentialized value is 17, a 0.25 probability that it is 16, and a 0.3 probability that it is 15. The confidentialized value is 15.

The key message of Figure 15.5 is that the model treats the unconfiden-

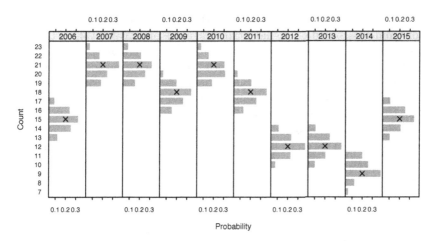

(a) Migration from South Region to Capital Region.

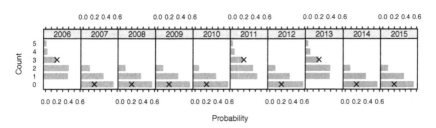

(b) Migration from South Region to West Region.

FIGURE 15.5: Posterior distributions for unconfidentialized migration counts. The counts are for females aged 20 moving out of South Region. Posterior distributions are shown for each year from 2006 to 2015. The × symbols indicate the confidentialized values.

tialized counts as uncertain. We do not pick out one set of numbers and treat that as our answer. Instead, we work with distributions.

There is, however, another subtle feature of Figure 15.5 that is very Bayesian. Looking closely at the distributions in the upper graph, we can see that they are not all symmetric around the confidentialized values. Instead, distributions near the top of the range are weighted toward lower values, while distributions near the bottom of the range are weighted towards higher values. The asymmetries are even stronger in the lower graph. None of the distributions based on confidentialized values of 0 place all their weight on 0. All of the distributions based on confidentialized values of 3 favour values 1 and 2.

In the terminology of Section 8.6, the model is shrinking the distributions towards a common mean. If a confidentialized value is unusually high, then the model places a higher-than-usual probability on the possibility that the value had been rounded up. If a confidentialized value is unusually low, then the model adjusts the probabilities in the opposite direction.

The shrinkage process leads to smoothing. The smoothing is clearest in the lower graph. The unconfidentialized values follow an straight line with occasional large deviations. The posterior distributions vary much less dramatically from year to year.

15.8   Results for Migration Rates

The posterior distributions for the migration rates, unlike the distributions for the migration counts, can be summarized nicely by credible intervals and posterior medians. We therefore return, in Figure 15.6, to our standard format.

The left-hand panels in Figure 15.6 show results from our main analysis in which we explicitly model the confidentialization process. For comparison, we also show, in the right-hand panels, results from an analysis with exactly the same system model, but where we ignore the fact that the data have been confidentialized. In other words, results in the left-hand panels were obtained using the framework of Part IV, while results in the right-hand panels were obtained using the framework of Part III.

Figure 15.6 also includes direct estimates, shown in black. Both sets of rates are smoother than the direct estimates. However, rates from the main analysis, with the model of the confidentialization process, smooth much more. The extra smoothing is particularly apparent in the lower panel, where cell counts are small.

Our main model calculates the rates jointly with the unconfidentialized counts. As we saw in Section 15.7, the model effectively smooths through the confidentialized counts to produce estimates of the unconfidentialized counts. It is therefore not surprising that the main model produces relatively smooth series for rates.

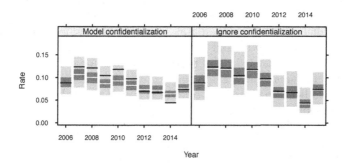

(a) Migration from South Region to Capital Region.

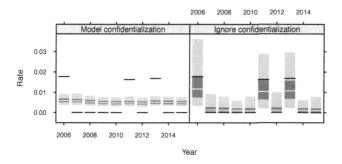

(b) Migration from South Region to West Region.

FIGURE 15.6: Estimated migration rates, by time, for females aged 20 migrating from South Region. The graphs on the left show results from the main analysis, with an explicit model of the confidential\izion process. The graphs on the right show what happens when we ignore the confidentialization process, and treat the confidentialized data as noise-free. The black lines show direct estimates, based on confidentialized data.

The smooth series of migration rates produced by the main model is much more plausible than the jagged series produced by the simpler model. There is no reason to expect underlying (super-population) migration rates to fluctuate wildly from year to year.

Aside from smoothing, another difference between the results from our main analysis and the one without the data model is that the credible intervals from the main analysis are much narrower. Why would this be?

In general, the value for standard deviation σ in a prior model such as that of Equation eq15.2 can be interpreted as a measure of the ability of the prior model to predict variation in the data. The lower the value for σ, the better the fit between prior model and data. The posterior median for σ from the main analysis is around 0.2. The posterior median for σ from the analysis with no data model is almost 1.5. Applying exactly the same system model to data that have been corrected for confidentialization and data that have not produces dramatically different results.

At first sight, these dramatic differences seems strange. Values that have been randomly rounded differ from original values by two at most, and are often identical. When the original values are in the tens, hundreds, or thousands, differences of 1 or 2 are indeed trivial. But when the original values are small, a difference of 1 or 2 can be large in relative terms. For instance, rounding a migration count from 1 to 3 leads to a direct estimate for the migration rate that is three times higher than the original.

15.9 Forecasting

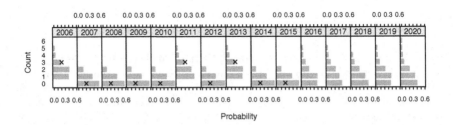

FIGURE 15.7: Estimated and forecasted unconfidentialized migration counts, for 20 year old females migrating between South Region and West Region. The bars show posterior distributions for the unconfidentialized counts, and the × symbols show the confidentialized counts.

To construct a forecast for the parameters of the system model, we follow the same steps that we did when forecasting model parameters in Part III (e.g. Sections 11.7 and 12.8). We generate the highest level of parameters, then generate the next level conditional on the highest level, and so on down to the migration rates.

To construct a forecast for the true, unconfidentialized migration counts, we continue the process one step further. We generate migration counts, conditional on the migration rates and exposures using Equation (15.1). Figure 15.7, for instance, shows estimates and forecasts for 20 year old females migrating from South Region to West Region.

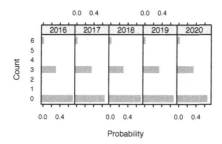

FIGURE 15.8: Forecasts of confidentialized migration counts, for 20 year old females migrating between South Region and West Region.

We can continue the forecasting process another step, and generate future confidentialized migration counts. To do this, we apply the random rounding rules to each draw from the posterior sample for the unconfidentialized counts. The result is a posterior sample for the confidentialized counts. Figure 15.8 shows forecasts of confidentialized counts for 20 year old females migrating from South Region to West Region.

15.10   References and Further Reading

The data on migration come from the table *Internal migration between regions by sex and age 1986-2016 - Division into municipalities as of 1 January 2017* on the Statistics Iceland website, downloaded on December 17, 2017. The original data are unconfidentialized: we randomly rounded the numbers to base 3 ourselves. We calculated exposures from data in table *Population by municipality, age and sex 1998-2017 - Division into municipalities as of 1 January 2017*, also on the Statistics Iceland website, downloaded on December 17, 2017.

The rules for random rounding to base 3 are described in Hundepool et al. (2012, p. 195).

16

Fertility in Cambodia

We would like to estimate age-specific birth rates in each of the 24 provinces of Cambodia. The Cambodian census is an obvious place to look for information. Cambodia conducts a census every 10 years or so, and the census contains a question on births. In 2008, the census yielded data on 179,000 births, which provides large cell sizes even after disaggregating by province and age of mother.

The 2008 Cambodian census happens, however, to be well-known example of how censuses in developing countries undercount births. The online demographic textbook *Tools for Demographic Estimation* states that "the results from the 2008 Census data suggest implausibly low levels of fertility in Cambodia ... It appears that only about half the births that occurred in the year before the census were reported to census enumerators."

Cambodia has had a number of Demographic and Health Surveys (DHS). These are standard international surveys, and produce data of high quality. The DHS for 2010, however, has a sample size only 100th as large as the 2008 Census. If we disaggregate DHS data on births in 2010 by province and 5-year age group, median cell size is only 9.

In this chapter, we show how the framework of Part IV can be used to estimate age-province-specific birth rates for Cambodia in 2010, in a way that capitalizes on the large sample size of the 2008 census and the accuracy of the 2010 DHS. We illustrate how to use information on demographic rates and coverage ratios, plus some demographic intuition, to iteratively check and improve our models. We begin, however, by reviewing the census and DHS data.

16.1 Data

Looking at Figure 16.1, it is clear why the authors of *Tools for Demographic Estimation* were concerned about under-reporting in the 2008 census. At the national level, fertility rates calculated from the census are only about half as high as the fertility rates calculated from the 2010 DHS. The undercount seems to be particularly pronounced in the 20s and early 30s, the ages of peak fertility. At ages 45–49, however, the census actually reports a higher birth rate than the DHS.

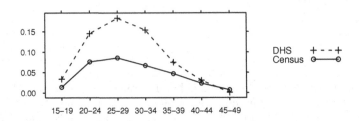

FIGURE 16.1: Direct estimates of age-specific fertility rates at the national level, based on the 2008 Census and 2010 Demographic and Health Survey.

Figure 16.2 shows fertility rates for provinces. The rates are calculated from census data. To save space, the figure only shows 8 of the 24 provinces. The provinces are ordered by the percent of the population below the poverty line. The poorest province is Preah Vihear, the next poorest is Kratie, and so on down to Phnom Penh Province, which contains the capital city.

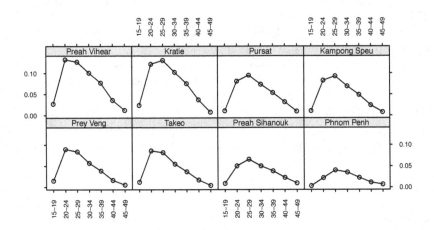

FIGURE 16.2: Direct estimates of age-specific fertility rates for eight selected provinces, based on the 2008 Census.

The poverty measures come from a United Nations Development Programme project which used small area estimation techniques to estimate local-level poverty rates. We use data for 2009. Phnom Penh Province has a poverty rate of only 0.2%. The next wealthiest province has a rate of 17.6%, and Preah Vihear has a rate of 43.1%. There is a clear relationship between provincial poverty rates and provincial fertility rates: poorer provinces have higher fertility.

Our exposure measure is the number of women counted in the 2008 Census, disaggregated by 5-year age group and province.

16.2 Overview of Model

Our overall model contains a system model to describe demographic behavior, and data models to describe the measurement processes (see Figure 16.3). The system model relates true birth counts to exposure. We treat exposure as known. The first data model describes the relationship between the true births counts and the census data. The second data model describes the relationship between the true births counts and the DHS data.

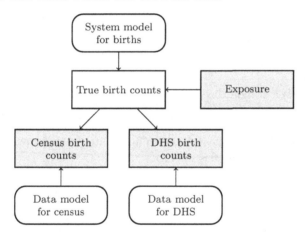

FIGURE 16.3: An overview of the model for births in Cambodia. Straight-edged rectangles represent demographic arrays, and rounded rectanges represent models. Gray shapes are observed; everything else must be inferred.

16.3 System Model

We use y_{ap} to denote the true, unobserved number of births to women aged a in province p. We assume that y_{ap} follows a Poisson distribution,

$$y_{ap} \sim \text{Poisson}(\gamma_{ap}^{\text{bth}} w_{ap}), \tag{16.1}$$

where w_{ap} is exposure, and γ_{ap}^{bth} is the birth rate for age group a in province p. Our prior model for the birth rates includes an age effect and a province effect,

$$\log \gamma_{ap}^{\text{bth}} = \text{N}(\beta^{\text{bth, 0}} + \beta_a^{\text{bth, age}} + \beta_p^{\text{bth, prov}}, \sigma_{\text{bth}}^2). \qquad (16.2)$$

We use a local trend model (Section 11.3.4) for the age effect, with the usual weakly informative priors on the standard deviations.

Our choice of prior for province effects is guided by Figure 16.2. The figure shows a clear relationship between poverty and fertility rates. Given this relationship, it would, in the terminology of Section 8.4, be inappropriate to treat the original provincial effects as exchangeable. Assuming exchangeability might, however, be appropriate if we adjust for differences in poverty levels first (see Section 8.5.2 for related discussion).

We assume that

$$\beta_p^{\text{bth, prov}} \sim \text{N}(\alpha + \delta z_p, \tau_{\text{bth}}^2), \qquad (16.3)$$

where z_p measures the poverty rate in province p. The parameter δ captures the relationship between poverty and fertility. We expect δ to be positive: if province p has high poverty (as measured by z_p) we expect it to have high fertility (as measured by $\beta_p^{\text{bth, prov}}$.)

We use a weakly informative prior on the slope parameter δ and a diffuse prior on the intercept parameter α.

Variable z_p is a standardized version of the original poverty variable: $z_p = (x_p - \bar{x})/(2s_x)$ where x_p is the original poverty variable, $\bar{x}$ is the mean of x_p, and s_x is the standard deviation. Standardizing in this way can stabilize the calculations, and make it easier to formulate priors for α and δ. We use priors $\alpha \sim \text{N}(0, 10^2)$ and $\delta \sim t_7^+(1)$.

16.4   Data Models

16.4.1   2008 Census

The data model for the census counts predicts the number of births reported in the census, given the true number of births. In contrast to the data model in the Iceland chapter, where we knew precisely how the data had been generated from the true counts, in this case we have to take some educated guesses.

We assume that births reported in the census, denoted by x_{ap}^{cen}, follow a Poisson distribution,

$$x_{ap}^{\text{cen}} \sim \text{Poisson}(\gamma_{ap}^{\text{cen}} y_{ap}). \qquad (16.4)$$

The parameter γ_{ap}^{cen} measures the expected number of births appearing in the census for each birth that actually occurs. We refer to it as a coverage ratio.

A value of 0.2, for instance, would imply that the census captured about 20% of births. As the *ap* subscripts indicate, we allow the coverage ratio to vary by age and province.

The γ_{ap}^{cen} in turn have a prior model,

$$\log \gamma_{ap}^{\text{cen}} = N(\beta^{\text{cen, 0}} + \beta_a^{\text{cen, age}} + \beta_p^{\text{cen, prov}}, \sigma_{\text{cen}}^2). \tag{16.5}$$

By including age effects $\beta_a^{\text{cen, age}}$ and province effects $\beta_p^{\text{cen, prov}}$ in the model, we are allowing for the possibility that coverage ratios vary systematically by age and province.

We use a local level model (Section 8.5.3) for the age effect. We expect neighbouring age groups to have similar coverage ratios, but do not expect a consistent trend upwards or downwards across age groups.

Our prior for the provincial effect is

$$\beta_p^{\text{cen, prov}} \sim N(0, \tau_{\text{cen}}^2). \tag{16.6}$$

We use an informative prior for τ_{cen}, the standard deviation in provincial effects: a half-t prior with scale 0.05. As discussed in Section 12.2, a difference of 0.05 on a log scale translates to a relative difference of 5 percentage points in ordinary units. And, as we saw in in Section 8.3.3, a normally distributed variable has an approximately 95% chance of being within two standard deviations of the mean. Therefore, in choosing a scale parameter of 0.05, we are saying that we expect 95% of provincial coverage ratios to be within 10 percentage points of the overall coverage ratio.

It is common to use informative priors like this when specifying data models. If the priors are too weak, then the model is unable to distinguish between genuine variation in rates or counts, and variation that is due merely to inconsistent coverage. If we have information on the quality of the data, then we should include it in the model. In a full scale analysis of the Cambodian fertility, we would not just take a guess at the appropriate value for τ_{cen}, but would talk to people who were knowledgeable about the data, or look for evidence from coverage surveys or field reports.

It is also important to remember that traditional demographic methods often make much stronger assumptions about errors in the data than the ones we make when setting up informative priors. In the section on the Cambodian census data, for instance, *Tools for Demographic Estimation* notes that most traditional methods for estimating fertility rates assume that coverage ratios are identical across age groups. In terms of our model, this is equivalent to assuming that every age effect $\beta_a^{\text{cen, age}}$ is exactly equal to 0.

16.4.2   2010 Demographic and Health Survey

Data for the 2010 Demographic and Health Survey were collected through a complicated process that involved dividing the Cambodian population into strata based on province and urban-rural residence, and then selecting communities and households to interview. Given this complexity, estimating the relationship between the DHS data and the underlying fertility rates is complicated and time-consuming. As a shortcut, we build on classical methods implemented in the R package **survey**.

Applying the classical methods yields a set of estimated age-specific birth counts x_a^{dhs} and associated standard errors s_a. Under the assumptions of the classical methods,

$$x_a^{\text{dhs}} \sim N(y_a, s_a^2). \tag{16.7}$$

The estimated number of births to women in age group a would, if the survey were repeated under identical conditions, equal, on average, the true number of births. Survey-to-survey variability is captured by variance term s_a^2.

Our data model for the DHS refers to age groups a, but not age-province combinations ap. In principle, we could disaggregate the DHS estimates along more dimensions than just age group. However, classical methods start to produce unreliable estimates once cell sizes become small. Classical estimates of the standard deviation terms are particularly problematic. Instead of trying to squeeze regional information out of the DHS data, we use it to calibrate age-specific estimates at the national level.

We use Equation (16.7) as our data model for the DHS data. By treating Equation (16.7) in this way, we are assuming that the survey performed as advertised: that its results really are unbiased, and that the s_a really do measure uncertainty properly. While this assumption in unlikely to be completely correct, it simplifies the analysis dramatically. It means that, like the data model for confidentialized data in Section 15.5, the data model for the DHS contains no unknown quantities.

Treating the DHS estimates as unbiased makes the most sense if we are thinking of the true birth counts y_a as occurring in the year 2010. In Equation (16.1), however, we use census counts from 2008 as our exposure measure. Our birth counts and exposure refer to different years. This inconsistency could be resolved by, for instance, adjusting the census counts for population growth between 2008 and 2010. For simplicity, however, we do not do that here.

16.5   Results

Having fitted the model, we obtain samples from the posterior distribution of true birth counts y_{ap}, birth rates γ_{ap}^{bth}, census coverage ratios γ_{ap}^{cen}, and all the other parameters.

Figure 16.4 shows estimates of birth rates for the eight selected provinces from Figure 16.2. The posterior distributions are all well above the direct estimates from the census, as we would hope. Some of the estimates, such as those for Preah Vihear, have considerable uncertainty. The estimates are higher in poorer provinces than in richer ones. Although we do not show it here, the pos-

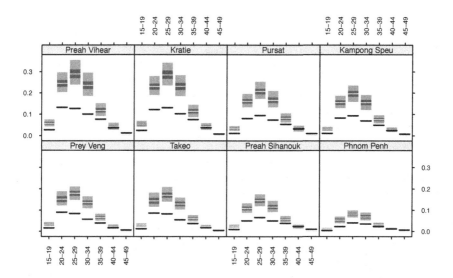

FIGURE 16.4: Estimates of age-specific fertility rates for eight selected provinces. The light gray bands represent 95% credible intervals, the dark gray bands represent 50% credible intervals, and the pale lines represent posterior medians. The black lines represent direct estimates from the census.

terior distribution for the δ parameter from Equation (16.3), which measures the relationship between poverty and fertility, is strongly positive.

Figure 16.5 shows estimates of national-level rates, which we obtain by summing up the provincial-level rates γ_{ap}^{bth}, weighted by the provincial-level exposures w_{ap}. The estimates are centered on the direct estimates from the DHS. These results are consistent with our assumption that the DHS direct estimates are unbiased.

We turn next to results from the data model for the census. Figure 16.6 shows estimates of coverage ratio γ_{ap}^{cen}. The dotted vertical lines indicate ratios of 1, where there is one census birth for every actual birth.

There is clearly a problem with the coverage ratios for births to women aged 45–49. Taken at face value, they imply that the census records 5–9 births to women in this age group for every birth that actually occurs. This is despite the fact that coverage ratios for other age groups are all at or below 1.

It turns out that the DHS estimates for births to women aged 45–49 are based on a total of three births. When cell sizes are this small, classical methods break down. In particular, the standard error s_a for ages 45–49 is suspiciously small. The strange results for 45–49 year olds probably owe more to malfunctioning DHS estimates than to problems in the census. We need to revise the model to take this into account.

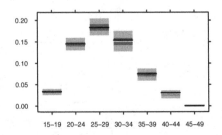

FIGURE 16.5: Estimates of age-specific fertility rates for the whole country. The black lines are direct estimates from the DHS.

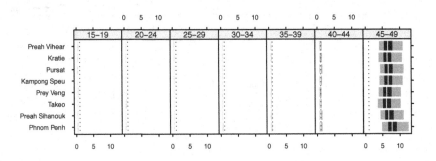

FIGURE 16.6: Estimates of coverage ratios for census data in eight selected provinces.

16.6   Revised Model

In our revised model, instead of using separate DHS estimates for age groups 40–44 and 45–49, we use a pooled estimate for age 40–49. This pooled age group has enough births for the classical estimation methods to function correctly.

In the data model, true birth counts y_{ap} for 40–44 year olds and 45–49 year olds are aggregated before they are compared with the DHS estimates, in the same way that true births across multiple provinces are aggregated before they are compared with DHS estimates. To obtain separate estimates for 40–44 year olds and 45–49 year olds, the revised model, in effect, exploits information on the ratio between the two age groups contained in the census data, and information on age-to-age trends contained in the prior for age effects.

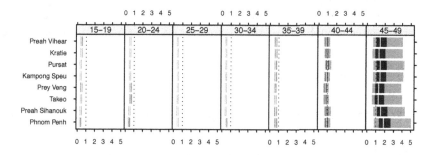

FIGURE 16.7: Estimates of census coverage ratios from the revised model.

Coverage ratios from the revised model are shown in Figure 16.7. Ratios for 45–49 year olds are still higher than for other age groups, but the differences are much less marked than they were with the previous model. It would be unwise to place complete faith in fertility estimates for this age group, but this may be as good as we can get with the data to hand.

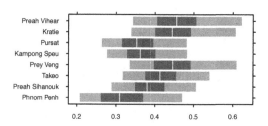

FIGURE 16.8: Estimates of census coverage ratios for 15–19 year olds from revised model.

Instead, we turn our attention to the coverage ratios for the youngest age group, 15–19 year olds. Looking closely at the left-most panel of Figure 16.7, the coverage ratio for 15–19 year olds seems to vary across provinces. Figure 16.8, which narrows in on 15–19 year olds, shows the variation more clearly.

Why would different provinces have different coverage ratios for 15–19 year olds? One possibility is that some provincial census organizations were more effective than others at encouraging participation by young people. An alternative explanation is suggested by the fact that

$$\text{coverage ratio} = \frac{\text{reported births}}{\text{actual births}}. \tag{16.8}$$

If the model overestimates actual births in a province, then the census coverage ratio in that province will appear unusually low. Conversely, if the model underestimates actual births, then the coverage ratio will appear unusually high.

The census coverage ratio for 15–19 year olds is highest in Preah Vihear, the poorest province, and lowest in Phnom Penh, the richest province. Maybe the model is systematically understating fertility for 15–19 year olds in poor provinces, and overestimating fertility for 15–19 year olds in rich provinces. Perhaps we need to revise the model again.

16.7   Final Model

We begin by examining the relationship between census coverage ratios for 15–19 year olds and poverty across all 24 provinces. Figure 16.9 shows posterior medians for census coverage ratios versus the percent of the population below the poverty line. Each dot represents one province.

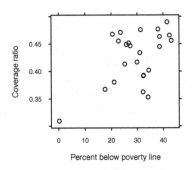

FIGURE 16.9: Estimates of coverage ratios for census data in eight selected provinces.

The dot at the bottom left of Figure 16.9 represents Phnom Penh. If we were to remove Phnom Penh from the graph, then there would essentially be no relationship between poverty and coverage ratios. We appear to need a special Phnom Penh effect, rather than a more general poverty-versus-coverage effect.

To do this we add a covariate to our prior model,

$$\log \gamma_{ap}^{\text{bth}} = \text{N}(\beta^{\text{bth, 0}} + \beta_a^{\text{bth, age}} + \beta_p^{\text{bth, prov}} + \eta v_{ap}, \sigma_{\text{bth}}^2). \qquad (16.9)$$

The variable v_{ap} takes a value of 1 for the cell where the age group a is 15–19 and the province p is Phnom Penh, and takes a value of 0 otherwise.

The parameter η measures the size of the "15–19 year olds in Phnom Penh" phenomenon.

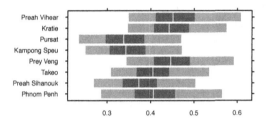

FIGURE 16.10: Estimates of census coverage ratios for 15–19 year olds from final model.

After fitting the model, we obtain the coverage ratios shown in Figure 16.10. Phnom Penh no longer stands out. Figure 16.11 compares age-specific fertility rates for Phnom Penh from the revised and final models. Fertility for 15–19 year olds is lower under the final model than the revised model, while fertility for older age groups is higher. The extra flexibility in the final model has produced a shift in the estimated age-pattern.

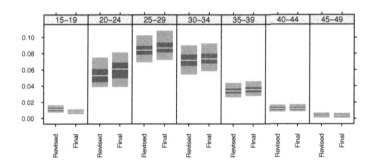

FIGURE 16.11: Estimates of age-specific fertility rates in Phnom Penh from revised and final models.

Looking back at Figure 16.7, coverage ratios for 20–24 year olds also vary across provinces, and might also also require further investigation. In a full scale study of Cambodian fertility, the revise-inspect-revise process might continue for several more rounds. We would also include some replicate data tests, and perhaps scrutinize rates for individual provinces more closely.

Rather than continue, however, we conclude the chapter by reviewing the results obtained at this point. Figure 16.12 shows age-specific fertility rates from the final model. As can be seen by comparing Figure 16.4, the changes we made to the original model have only had a minor effect on the overall results. This is a little reassuring. A model that gives dramatically different results in response to small changes in specification cannot really be trusted.

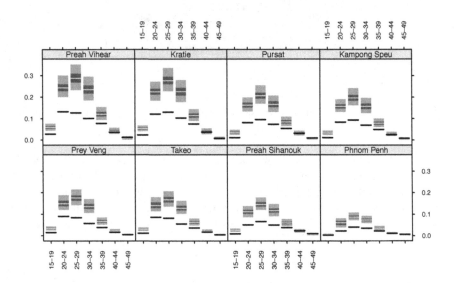

FIGURE 16.12: Estimates of age-specific fertility rates from the final model.

In addition to examining fertility rates for individual age groups, it is useful to have a summary measure that combines rates across age groups. Demographers' standard summary measure for fertility is the 'total fertility rate' (TFR). The TFR is the sum of the age-specific rates at each reproductive age. It equals the number of births the average woman would have over her lifetime if she survived through the reproductive ages, and if current age-specific fertility rates were to persist indefinitely.

Figure 16.13 shows total fertility rates for the Cambodian provinces. Although there is considerable uncertainty about the provincial rates, Preah Vihear and Kratie appear to have TFRs about 4 times higher than those in Phnom Penh.

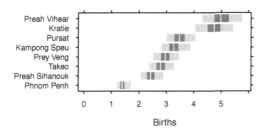

FIGURE 16.13: Estimates of total fertility rates for eight selected provinces.

16.8   References and Further Reading

The Demographic and Health Survey data are from National Institute of Statistics and ICF Macro (2011b). The 2010 survey is described in National Institute of Statistics and ICF Macro (2011a). We downloaded the census data from the *Cambodia General Population Census* database on the Latin American and Caribbean Demographic Center (CELADE) website, on November 21, 2017. The provincial poverty rate data come from Asian Development Bank (2014).

Moultrie et al. (2013) is an authoritative guide to traditional demographic methods for estimation from deficient data, and available at *demographicestimation.iussp.org*. The section on the 2008 Cambodia Census was written by Tom Moultrie.

Alkema et al. (2012) and Alexander and Alkema (2018) use Bayesian methods to combine multiple data sources and estimate fertility rates and mortality rates. Bryant and Graham (2015) show how coverage ratios can be used to diagnose problems in a model from the framework of Part IV, and discuss how variance parameters from data models can be used as measures of reliability.

Part V

Inferring Demographic Accounts

17

Inferring Demographic Accounts

Models for inferring demographic accounts are much larger and more complicated than models for inferring single demographic arrays. However, as we will see in Part V, they are composed of essentially the same elements as models for single arrays, combined in essentially the same way.

17.1 Summary of Our Approach

The framework of Part V is summarized in Figure 17.1. The components are:

Demographic account $Y_1, \cdots Y_K$. Array Y_1 holds population counts, and arrays $Y_2, \cdots, Y_K$ hold component counts, such as births and deaths. They are linked by accounting identities. An account must always have a population series, but all other series are optional. The account is not observed directly.

Arrays $\gamma_1, \ldots, \gamma_K$ of rates or means. Super-population rates or means appearing in the models for $Y_1, \cdots Y_K$. Rates are used in Poisson models in the rates-exposure form, and means are used in Poisson models in the counts form or in normal models (which are used with net migration). The two forms of Poisson models are discussed in Section 8.3.1.

Vectors $\phi_1, \cdots, \phi_K$ of parameters from prior models. Equivalent to ϕ in Parts III and IV.

Datasets $X_1, \cdots, X_M$. Identical to $X_1, \cdots, X_M$ in Part IV, except that they are restricted to arrays of counts.

Vectors $\Omega_1, \cdots, \Omega_M$ of parameters from data models. Identical to $\Omega_1, \cdots, \Omega_M$ in Part IV.

In contrast to the models of Parts III and IV, the models of Part V do not include a separate exposure term. Instead, exposures are calculated internally, using values from the population series.

Altogether, there are M datasets, each of which is an array of counts. Each dataset is associated with a single demographic series. However, a demographic

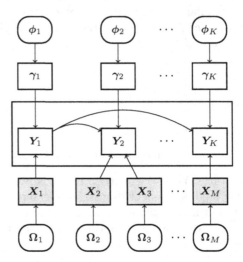

FIGURE 17.1: Inferring a demographic account, and associated rates or means, from one or more datasets. Y_1 is an array of population counts, and $Y_2, \cdots, Y_K$ are arrays of component counts. As indicated by the big rectangle, the arrays together form a demographic account. $\gamma_1, \gamma_2, \cdots, \gamma_K$ are arrays of rates or means, and $\phi_1, \phi_2, \cdots, \phi_K$ hold parameters from the associated prior models. Arrays $X_1, X_2, \cdots, X_M$ are datasets, and $\Omega_1, \Omega_2, \cdots, \Omega_M$ are parameter values from the associated data models. The only elements that are observed are $X_1, X_2, \cdots, X_M$.

series can be associated with any number of datasets. A demographic series can even have no data series, implying that there are no direct measures for the series.

The models for each of the demographic series $Y_1, Y_2, \cdots, Y_K$ are set up independently of each other. However, because the series are linked together by the demographic accounting identities, they are necessarily estimated as a group. Changing a value in one series requires changing values in at least one other series, to preserve the accounting identities. This in turn can affect rates and probabilities elsewhere. For instance, a smaller number of deaths implies a larger population, which implies more people exposed to the risk of having a birth or migrating.

If data for one demographic series are more comprehensive and accurate than data for another series, then the series with the superior data will tend to be estimated more precisely. If a country has good vital registration data on births and deaths, for example, but poor data on migration, then birth rates and death rates for that country will have relatively narrow credible intervals, while migration rates will have relatively wide intervals.

Some demographic series may not be measured at all. Many countries, for instance, collect no data on emigration. Estimation proceeds as normal in such cases, though the credible intervals are inevitably wider than they would have been if data had been available.

A distinctive feature of the models of Part V is that they treat all demographic series symmetrically. All series have γ_k, and priors governing the γ_k, and all series can have associated datasets. Traditional methods for estimating historical accounts do not generally treat series symmetrically. Instead, one series—typically population or emigration—is not modeled directly, but is instead derived by applying the demographic accounting identities to values for the remaining series.

As with the models for parts III and IV, the models of Part V can be used for forecasting. We simply extend the Y_k, γ_k, and ϕ_k into future periods, ignoring the fact that no data are available for these periods.

The models of Part V produce voluminous output, with estimates for each of the demographic series, for the parameters governing each series, and for the data models relating the series to the available data. As usual, all these estimates consist of draws from the posterior distribution. Each draw for the various demographic series is internally consistent, in that the demographic series conform to the accounting identities.

Estimates from different parts of the model are all connected with one another, in the sense that changes in one part imply changes in all the others. Consider, for instance, the typical series of changes that would occur after we expanded the model by adding a new dataset with high-quality measures of population size.

- Estimates of population counts, and the associated super-population values, shift towards their true values and become more precise.

- Exposure measures shift towards their true values and become more precise.

- Estimates of series such as birth counts and death counts that are modeled using exposures are affected, as are the associated super-population values.

- Estimates of the remaining series are affected, via the accounting identities.

- Estimates of super-population parameters in the data models are affected, since estimates of the true counts have changed.

Example 17.1. Figure 17.2 illustrates a hypothetical example of a demographic account. The account contains a single dimension, time. The model for population is

$$y_t^{\text{popn}} \sim \text{Poisson}(\gamma_t^{\text{popn}}) \tag{17.1}$$

$$\log \gamma_t^{\text{popn}} \sim \text{N}(\mu_{\text{popn}}, \sigma_{\text{popn}}^2) \tag{17.2}$$

$$\mu_{\text{popn}} \sim \text{N}(0, 1) \tag{17.3}$$

$$\sigma_{\text{popn}} \sim t_7^+(1), \tag{17.4}$$

where y_t^{popn} is the true number of people at time t. The model for births is

$$y_t^{\text{bth}} \sim \text{Poisson}(\gamma_t^{\text{bth}} w_t) \tag{17.5}$$

$$\log \gamma_t^{\text{bth}} \sim \text{N}(\mu_{\text{bth}}, \sigma_{\text{bth}}^2) \tag{17.6}$$

$$\mu_{\text{bth}} \sim \text{N}(0, 1) \tag{17.7}$$

$$\sigma_{\text{bth}} \sim t_7^+(1), \tag{17.8}$$

where y_t^{bth} is the true number of births occurring during the period between exact times $t - 1$ and t, and w_t is an exposure term calculated using $w_t = (y_{t-1}^{\text{popn}} + y_t^{\text{popn}})/2$. (The use of population counts to approximate exposure is discussed in Section 4.6.) The model for deaths has the same structure as the model for births,

$$y_t^{\text{dth}} \sim \text{Poisson}(\gamma_t^{\text{dth}} w_t) \tag{17.9}$$

$$\log \gamma_t^{\text{dth}} \sim \text{N}(\mu_{\text{dth}}, \sigma_{\text{dth}}^2) \tag{17.10}$$

$$\mu_{\text{dth}} \sim \text{N}(0, 1) \tag{17.11}$$

$$\sigma_{\text{dth}} \sim t_7^+(1). \tag{17.12}$$

The first dataset is the census, which measures population size in the year 2000. The data model is

$$x_t^{\text{cen}} \sim \text{Binomial}(y_t^{\text{popn}}, \pi_{\text{cen}}), \tag{17.13}$$

where π_{cen} is the probability that a person is enumerated by the census.

The second dataset is enrolment data from the health system. The data model for the enrolment data is

$$x_t^{\text{hlh}} \sim \text{Poisson}(\lambda_{\text{hlh}} y_t^{\text{popn}}). \tag{17.14}$$

μ:	?
σ:	?

Parameters $\boldsymbol{\phi}_{\mathrm{popn}}$

μ:	?
σ:	?

Parameters $\boldsymbol{\phi}_{\mathrm{bth}}$

μ:	?
σ:	?

Parameters $\boldsymbol{\phi}_{\mathrm{dth}}$

2000	?
2001	?
2002	?
2003	?

Expected popn $\boldsymbol{\gamma}_{\mathrm{popn}}$

2001	?
2002	?
2003	?

Birth rates $\boldsymbol{\gamma}_{\mathrm{bth}}$

2001	?
2002	?
2003	?

Death rates $\boldsymbol{\gamma}_{\mathrm{dth}}$

2000	?
2001	?
2002	?
2003	?

Population $\boldsymbol{Y}_{\mathrm{popn}}$

2001	?
2002	?
2003	?

Births $\boldsymbol{Y}_{\mathrm{bth}}$

2001	?
2002	?
2003	?

Deaths $\boldsymbol{Y}_{\mathrm{dth}}$

2000	96

Census $\boldsymbol{X}_{\mathrm{cen}}$

2000	105
2001	121
2002	111
2003	119

Health roll $\boldsymbol{X}_{\mathrm{hlh}}$

2001	18
2002	14
2003	21

Reg births $\boldsymbol{X}_{\mathrm{rbth}}$

2001	9
2002	17
2003	12

Reg deaths $\boldsymbol{X}_{\mathrm{rdth}}$

π:	?

Parameters $\boldsymbol{\Omega}_{\mathrm{cen}}$

λ:	?

Parameters $\boldsymbol{\Omega}_{\mathrm{hlh}}$

π:	?

Parameters $\boldsymbol{\Omega}_{\mathrm{rbth}}$

π:	?

Parameters $\boldsymbol{\Omega}_{\mathrm{rdth}}$

FIGURE 17.2: A simple example of a model using the framework of Part V.

The parameter λ_{hlh} measures the average number of people on the roll, for each person in the population.

The third dataset is registered births, which has data model

$$x_t^{\text{rbth}} \sim \text{Binomial}(y_t^{\text{bth}}, \pi_{\text{rbth}}), \qquad (17.15)$$

and the fourth is, registered deaths, which has data model

$$x_t^{\text{rdth}} \sim \text{Binomial}(y_t^{\text{dth}}, \pi_{\text{rdth}}). \qquad (17.16)$$

□

17.2   Applications

The classic application of demographic accounts is estimating and forecasting counts of population, births, deaths, and migration, for regions within a country. This remains an important application, given the huge demand for information on population stocks and flows at the local level. However, the framework is extremely flexible, and can be applied to many other problems.

We can, for instance, estimate a demographic account for a country that includes ethnicity along with age, sex, and region. We can model people moving between ethnicities in the same way that we model them moving between regions. We do not necessarily have to have direct measures of these moves, since we can infer them from the accounting identities. Estimates of ethnic identity change would come with measures of uncertainty. Forecasts produced within the system would incorporate these uncertainties.

There is no need to confine ourselves to national populations. We could, for instance, define our population of interest to be people with medical qualifications, and disaggregate along dimensions such as age, sex, speciality, and practicing versus non-practicing. Some subpopulations, such as people currently practicing would be relatively easy to estimate, while others, such as those who had left the labor market, would be harder to estimate.

Super-population quantities, rather than finite-population quantities, could be our main focus. We could, for instance, use the methods of Part V to obtain consistent integrated estimates of fertility rates, mortality rates, and migration rates.

As with the methods of Part IV, we might even be seeking insights into the data sources, rather than the demographic system itself. We might want to compare the completeness of birth registrations and death registrations, for instance, or assess how ethnicity is recorded within different administrative data sources.

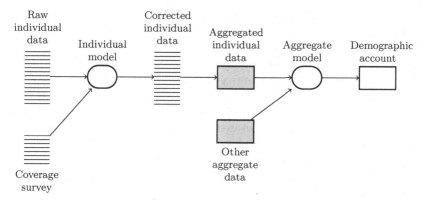

FIGURE 17.3: Combining individual-level approaches to population estimation with Bayesian demographic accounts.

17.3 Demographic Accounts in Official Statistical Systems

As we discussed in Section 6.2, national statistical agencies in many countries are developing statistical systems based on linked administrative datasets. If these new systems were to work perfectly, then a demographer who wanted to know how many births there were in the country, for instance, could read the numbers straight off the individual datasets, rather than bothering with Bayesian demographic accounts.

It seems likely to us that individual-level datasets will indeed become increasingly important sources of demographic information in countries that have the required statistical infrastructure. But even in these countries, there will still be a place for demographic accounts.

In most countries, it will be difficult to push the individual-level datasets far back into the past. This is partly because many administrative data sources are new, or do not have the required quality in earlier periods. It is also because most individual-level approaches include some sort of coverage survey, and these cannot be carried out retrospectively.

A lack of long time series is a serious deficiency. Long time series make it possible to discern trends, and to test theories. And, as we have seen in earlier chapters, constructing a forecast without a long time series of data to draw on is difficult.

Compared with individual-level approaches, Bayesian demographic accounts have modest data requirements. Individual-level approaches require individual-level datasets. Bayesian demographic accounts only require tabulations containing a few demographic variables, and different tabulations can contain different variables, or cover different periods. This makes it much easier to extend estimates back into the pre-digital era.

The modest data requirements also allow Bayesian demographic accounts to exploit data sources that individual-level methods cannot. Bayesian demographic accounts require only that counts in the data sources are correlated with the demographic series of interest. Data on cell phone use, electricity connections, or building consents could all, for instance, be used as proxies for population growth in a demographic account. It is not clear, however, how these sources be used in the construction of an individual-level database.

A further strength of Bayesisan demographic accounts is that they include, via the system models, tests of demographic plausibility. When they have sufficient data to estimate long-run trends in fertility and mortality, for instance, they should automatically downweight outcomes that would imply dramatic breaks from these trends.

In countries with advanced statistical systems, we see demographic accounts as complementing individual-level approaches. The relationship between the two is depicted in Figure 17.3. Statisticians apply models and coverage surveys to the raw individual-level datasets, to construct corrected individual-level datasets. Even with the best data and best methods, however, these individual-level datasets will not be perfect. The outputs from the individual-level modeling can be used as inputs to Bayesian demographic accounts. The result, hopefully, would be aggregate-level demographic estimates that were better than ones read straight from individual-level datasets.

Of course, in much of the world, the idea of assembling accurate individual-level information on the entire population is still a distant dream. For these countries, demographic accounts are, for the time being, a more realistic goal.

17.4   References and Further Reading

Rees (1985) discusses the difficulties of estimating demographic accounts. Tuljapurkar (2013) is a mathematical study of population dynamics that includes random events and random demographic rates. Bryant and Graham (2013) is an earlier, less general, version of the demographic accounting framework here.

18

Population in New Zealand

In this chapter we use Bayesian demographic accounts to study the growth of the New Zealand population. We begin with the simplest possible demographic system, with no age, sex, or any other way of classifying the population. We then extend the analysis by adding a region dimension. The two systems, which we refer to as the national system and regional system, are summarized in Table 18.1.

TABLE 18.1

Specifications for the two demographic systems modeled in Chapter 18

	National system	Regional system
Membership	Usual resident	Usual resident
Classification	None	Region
Entries	Births, immigration	Births, immigration
Exits	Deaths, emigration	Deaths, emigration
Movements	None	Internal migration

A person is a member of either of these systems if he or she is a "usual resident" of New Zealand, meaning that the person considers New Zealand to be his or her main country of residence. A person enters the usually-resident population of New Zealand by being born in New Zealand, or by immigrating to New Zealand, and exits the usually-resident population by dying or emigrating. Because the national system puts everyone into the same category, it does not allow for changes in status. The regional system, in contrast, distinguishes between regions of residence, which people can change via internal migration.

Compared with most countries, New Zealand has abundant, high-quality demographic data. Even in New Zealand, however, the data are neither complete nor fully accurate. To estimate our two demographic systems, we need to build models of the data sources, and then compare the data sources against each other and against our prior expectations about demographic trends.

18.1   Input Data for the National Demographic Account

We analyze the national demographic system by setting up a national demographic account. Table 18.2 summarizes the data sources we use for the account. "Census-based population estimates" are population estimates for years during which population censuses were conducted. New Zealand normally holds a census every 5 years, but the 2011 census was delayed until 2013 because of an earthquake. Census-based estimates are constructed by taking the raw census counts and adjusting for people missed from or mistakenly included in the census, as well as a number of smaller adjustments. The main source of information for the adjustments is a survey of census coverage carried out immediately after the census. Statistics New Zealand (Stats NZ) has constructed approximate credible intervals for the 2013 estimates. These credible intervals imply that, at the national level, the estimated value for population should be less than 1% away from the true value.

TABLE 18.2
Data sources for national demographic account

Data source	Series	Accuracy
Census-based popn estimates	Population	Good
Admin-based popn estimates	Population	Good
Registered births	Births	Excellent
Registered deaths	Deaths	Excellent
International arrivals	Immigration	Moderate
International departures	Emigration	Moderate

The "admin-based population estimates" in Table 18.2 are constructed from administrative datasets supplied to Stats NZ by other government departments. The administrative datasets include, for instance, tax data, schools enrollments, and health system enrollments. Stats NZ uses information such as name and date of birth to link individuals across multiple datasets. (Access to the individual-level data is strictly controlled, to protect confidentiality.) By keeping track of births, deaths, immigrations, and emigrations, and by applying "sign of life" checks where an individual has to show up occasionally in at least one dataset, Stats NZ can build up an approximate database of the usually-resident population. Based on comparing results from the two series, Stats NZ analysts suspect that the accuracy of the admin-based population estimates may be approaching that of the conventional census-year population estimates. The admin-based estimates are available annually, but only from 2007.

Births and deaths data in New Zealand come from an efficient registration system, and are of high quality. The most notable problem with the data is that there are sometimes delays in the reporting of births.

At first sight, it might seem that data on international arrivals and departures would be of similar high quality, given that New Zealand is an island with an efficient administrative system. New Zealand does indeed record virtually every entry to and exit from the country. The problem is distinguishing between movements that count as changes of usual residence and movements that are only temporary. The vast majority of movements are temporary movements, made by people on holidays, business trips, or family visits. Accurately identifying the small proportion of permanent moves can be difficult. The arrivals and departures data that we use in this chapter rely on the answers that people give on entering and leaving the country, when they are asked if they are moving permanently.

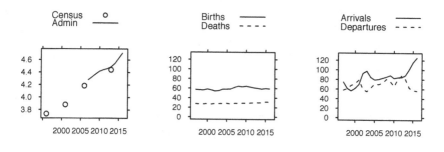

(a) Population estimates (millions) based on census and administrative data.

(b) Registered births and registered deaths (thousands)

(c) International arrivals and international departures (thousands)

FIGURE 18.1: Input data for the national demographic account. The census-based population estimates are based on census data for 1996, 2001, 2006, and 2013, adjusted for coverage errors. The administrative estimates are based on linked government datasets. Both sets of estimates were constructed by Stats NZ. Births and deaths data are from the vital registration system.

National-level counts from the various data sources are graphed in Figure 18.1. Neither the census-based population estimates nor the admin-based ones cover every year of the estimation period. Where the two estimates overlap, in 2013, they are close but not identical. Arrivals and departures fluctuate far more than registered births and deaths. Beginning around 2012, reported net migration, that is, arrivals minus departures, climbed to reach record levels.

18.2   Model for National Demographic Account

18.2.1   Overview

Even though our first demographic system makes no distinctions within the New Zealand population, the fact that there are multiple sources of entry and exit, and multiple data sources, means the associated accounting model is still complicated. Figure 18.2 summarizes the model.

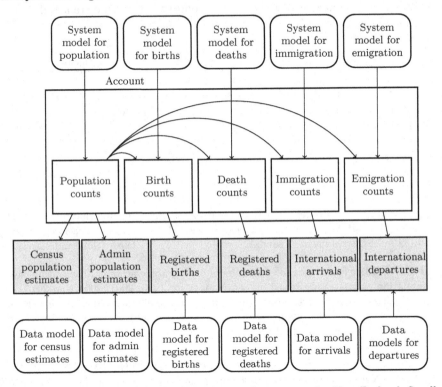

FIGURE 18.2: Overview of national demographic account for New Zealand. Small straight-edged rectangles represent demographic arrays, and rounded rectangles represent models. The gray rectangles are observed; everything else is inferred. The big rectangle encloses the demographic account.

The core of the model is the demographic account, marked out in Figure 18.2 by the big rectangle. As discussed in Sections 1.2 and 17.1, exposures for counts of births, deaths, immigration, and emigration are calculated internally using population counts. Our aim is to infer the various demographic series making up the account. To do this we set up one system model for each series: a system model for population counts, a system model for birth counts, and so on. We have two datasets for population, and one for each of the remaining series. Each dataset has an associated data model, describing the relationship between the true counts and the contents of the dataset.

18.2.2 Account

The format of the demographic account is illustrated in Figure 18.3. We aim to estimate the population at 21 time points, and estimate entries and exits for the 20 periods enclosed by these time points. As usual with Bayesian modeling, we aim to estimate the full posterior distribution for the accounts, rather than just point estimates. We will be producing hundreds or thousands of versions of Figure 18.3, each of which is a draw from the posterior distribution.

1996	1997	...	2016
3.73	3.77	...	4.66

(a) Population (millions)

	1997	1998	...	2016
Births	0.06	0.06	...	0.06
Deaths	0.03	0.06	...	0.03
Immigration	0.07	0.06	...	0.12
Emigration	0.07	0.06	...	0.06

(b) Components (millions)

FIGURE 18.3: Format of the national demographic account. The account uses "June years". The population counts refer to June 30, 1996, June 30, 1997, and so on. The counts for components refer to the period between July 1, 1996 and June 30, 1997, between July 1, 1997 and June 30, 1998, and so on.

18.2.3 System Models

The system models describe regularities in the true counts of population and events. The national demographic system is so simple that the only only regularity to be described is variation over time. We expect that the counts will evolve smoothly over time, with the possibility of occasional bumps.

We model population counts with

$$y_t^{\text{popn}} \sim \text{Poisson}(\gamma_t^{\text{popn}}) \tag{18.1}$$

$$\log \gamma_t^{\text{popn}} \sim \text{N}(\beta^{0,\,\text{popn}} + \beta_t^{\text{time, popn}}, \sigma_{\text{popn}}^2). \tag{18.2}$$

Equation (18.1) differs from the Poisson models used in previous chapters (e.g., Section 12.6.1 or Section 15.4) in that it does not include an exposure term. As we discuss in Section 8.3.1, models of population counts do not generally have exposure terms. The γ_t^{popn} in Equations (18.1) and (18.2) is an expected count rather than a rate.

The prior model for population consists of an intercept $\beta^{0,\,\text{popn}}$ and a time effect $\beta_t^{\text{time, popn}}$. The intercept governs the overall level, i.e. the average population size for the whole period, and the time effect governs change over time. We use a local trend model (Section 11.3.4) for the time effect. The local trend model allows for consistent trends upwards or downwards.

The standard deviation terms in the local trend model all have half-t priors, with scales of 0.05. The value of 0.05 prior is lower, and the prior correspondingly stronger, than our default value of 1. Demographic accounting models of Part V can be slow to converge. Using informative priors can speed up convergence. Taking the trouble to find an appropriate value for scale terms can therefore be worthwhile. Moreover, a scale of 0.05 allows for year-on-year changes in population of 10% or so, which is a bigger change than we would expect to see in practice. Even if our prior is stronger than our default one, it is still not as strong as it might be.

The exact value of 0.05 for the scale is, of course, somewhat arbitrary. Why not 0.04, or 0.075? In fact, we assume that using the values 0.04 or 0.075 would lead to very similar answer to using 0.05. Assumptions about sensitivity to alternative values can be checked, as we do in Section 18.4. We are not trying to choose some sort of optimal value for the scale, but merely to get the order of magnitude right: 0.05 rather than 0.005 or 0.5.

The system model for birth counts is

$$y_t^{\text{bth}} \sim \text{Poisson}(\gamma_t^{\text{bth}} w_t) \tag{18.3}$$

$$\log \gamma_t^{\text{bth}} \sim \text{N}(\beta^{0,\,\text{bth}} + \beta_t^{\text{time, bth}}, \sigma_{\text{bth}}^2). \tag{18.4}$$

The w_t in Equation (18.3) is an exposure, calculated as $w_t = (y_t^{\text{popn}} + y_{t+1}^{\text{popn}})/2$. (The use of population counts to approximate exposure is discussed in Section 4.6.) In the models of Part III and IV, exposures are fixed and known. Here, however, they are calculated as part of the overall estimation process.

The time effect in Equation 18.4 has a local level model (Section 8.5.3). We use a local level model, rather than a local trend model, because birth counts, unlike population counts, have not been following any clear trend upwards or downwards in New Zealand. We use the same half-t prior with scale 0.05 for standard deviations as we do in the model for population.

The system models for deaths, immigration, and emigration all have the same structure as the system model for births. The system model for deaths uses 0.05 as the scale for the half-t priors, while the system models for immigration and emigration use 0.2. Immigration and emigration vary much more from year to year than births and deaths.

18.2.4 Data Models

To represent the accuracy of the census-based population estimates, we use a normal distribution,

$$x_t^{\text{cen}} \sim \text{N}(y_t^{\text{popn}}, s_t^2). \tag{18.5}$$

This is a similar data model to the one we use for birth counts from the Cambodian Demographic and Health Survey in Section 16.4.2. We choose values for s_t that imply that the census count x_t^{cen} has an approximately 95%

chance of being within 0.5% of the true population count y_t^{popn}. As discussed in Section 8.3.3, a normally distributed variable has an approximately 95% chance of being within two standard deviations of the mean. Hence x_t^{cen} has an approximately 95% chance of being within $2s_t$ of the true population count y_t^{popn}. We set s_t to $0.0025 \times x_t^{\text{cen}}$, such that $2s_t = 0.005 \times x_t^{\text{cen}} \approx 0.005 \times y_t^{\text{popn}}$.

The admin-based population estimates receive a less precise data model,

$$x_t^{\text{adm}} \sim \text{Poisson}(\gamma_t^{\text{adm}} y_t^{\text{popn}}), \tag{18.6}$$

$$\log \gamma_t^{\text{adm}} \sim \text{N}(\mu^{\text{adm}}, \sigma_{adm}^2). \tag{18.7}$$

The γ_t^{adm} in Equation (18.6) measures the expected number of people reported in the admin-based population estimate for each person in the true population. Values greater than 1 imply over-coverage, and values less than 1 imply under-coverage. As the t subscript indicates, we allow for the possibility that coverage ratios vary over time.

The mean parameter μ^{adm} in Equation (18.6) controls the average coverage ratio over the whole period. We set

$$\mu^{\text{adm}} \sim \text{N}(0, 0.025^2), \tag{18.8}$$

which implies that we expect μ^{adm} to be somewhere in the range (-0.05, 0.05). The range (-0.05, 0.05) on the log scale translates to (0.95, 1.05) in ordinary units. Our prior for μ^{adm} therefore implies that we expect the distribution of the coverage ratios γ_t^{adm} to be centered on a point somewhere between 0.95 and 1.05. We expect any systematic biases in the dataset to be relatively small.

The individual values of $\log \gamma_t^{\text{adm}}$ in Equation (18.7) are drawn from a distribution centered on μ^{adm}, rather than equalling μ^{adm} exactly. In other words, we allow for the possibility that coverage ratios vary idiosyncratically from their overall average. The amount of variation is governed by parameter σ_{adm}, which has a half-t distribution with a scale of 0.025. Again, we are expecting idiosyncratic errors of a few percentage points at most.

With births and deaths, we use a model specifically designed for representing accurate data sources, which we refer to as a Poisson-binomial model. The model assumes that a reported count may have small errors, but that these errors are equally likely to be positive and negative.

A substantive interpretation of a Poisson-binomial distribution is as follows. Suppose that the data says there are X people or events for a given cell. The number X is the sum of (i) real people or events that are counted correctly, and (ii) people or events that are overcounted or wrongly included in the data. The Poisson-binomial model assumes that people or events missing from (i) are, on average, compensated for by people or events appearing in (ii). Therefore the reported count is on average equal to the true count.

Let n be the true count, and p the probability that a real person or event is reported. We use

$$y \sim \text{Poisson-binomial}(p, n)$$

to signify that y is drawn from a Poisson-binomial distribution. Parameter p is supplied by the user. The closer p is to 1, the more accurate the data source is.

Let U be the number of real people or events that are correctly counted in the data. Then U follows a binomial distribution: $U \sim$ binomial(n, p). Let V be the number of people or events that are overcounted or wrongly included. Suppose that V independently follows a Poisson distribution, with rate $(1-p)$ and exposure equal to the true count n. Then, in our terminology, $X = U + V$ has a Poisson-binomial distribution with parameters p and n. The mean of variable X is $pn + (1-p)n = n$, equal to the true count. The variance of X is $np(1-p) + n(1-p) = n(1-p^2)$. A Poisson variable with mean n would have variance n. Therefore, when p is close to 1, the variance of a Poisson binomial variable with parameters p and n is substantially smaller than that of a Poisson variable with parameter n.

In the New Zealand demographic account we set p to 0.98 for births and 0.99 for deaths, that is,

$$x_t^{\text{bth}} \sim \text{Poisson-binomial}(0.98, y_t^{\text{bth}}) \qquad (18.9)$$

$$x_t^{\text{dth}} \sim \text{Poisson-binomial}(0.99, y_t^{\text{dth}}). \qquad (18.10)$$

We chose these values after discussing the magnitude of likely errors with Stats NZ staff. It turns out, however, that the model results are fairly insensitive to the particular choice of p.

We model international arrivals and departures using Poisson models with the same likelihood and priors as the model for the admin-based population estimates. The likelihoods are

$$x_t^{\text{arr}} \sim \text{Poisson}(\gamma_t^{\text{arr}} y_t^{\text{arr}}), \qquad (18.11)$$

$$x_t^{\text{dep}} \sim \text{Poisson}(\gamma_t^{\text{dep}} y_t^{\text{dep}}). \qquad (18.12)$$

18.2.5 Estimation

We estimate the model using our R packages. The model is built piece by piece: there are functions to create priors, functions to create system models and data models, and a function to tie all the models and data together. As with all chapters in Parts III–V, code to run the models is available on the website for the book.

The estimation function iterates through each component of the model, updating that component, conditional on current estimates for the other components. It updates birth counts conditional on everything else, then death counts conditional on everything else, and so on through the account, the system models, and the data models.

Depending on the exact specification, the national account can require hundreds of thousands of iterations before it reaches complete convergence. This is a lot of iterations, given that the number of cells being estimated is small. The reason for the slow convergence is that the model has trouble separately identifying immigration and emigration. The data models for arrivals and departures are relatively weak.

Often, when data models are weak, the demographic accounting identities can provide extra constraints, which help restrict the range of estimates. But in this case, the constraints are themselves weak. Within our demographic system,

$$\begin{matrix} \text{Population} \\ \text{at end of} \\ \text{period} \end{matrix} = \begin{matrix} \text{population} \\ \text{at start} \\ \text{of period} \end{matrix} + \text{births} - \text{deaths} + \begin{matrix} \text{immi-} \\ \text{gration} \end{matrix} - \begin{matrix} \text{emigra-} \\ \text{tion.} \end{matrix}$$

Population, birth, and death counts together constrain the difference between immigration and emigration, but not the overall levels. If we have a set of immigration and emigration values that satisfy the accounting constraints, and we add 100,000 to all values for immigration and all values for emigration, then the new set of values will also satisfy the accounting constraints, since the extra immigrations and emigrations cancel out. The system models and data models do in fact provide enough information for the model to infer posterior distributions for immigration and emigration. But the signal is weak, and it takes a long time to reconstruct it.

18.3   Results for the National Demographic Account

The estimates for population counts are shown in Figure 18.4, together with the census-based and admin-based population estimates.

Looking closely at Figure 18.4, we can see that estimates are most precise—the credible intervals are narrowest—in years with a census, and least precise in years without a census. Figure 18.5 brings out the differences more clearly, by showing the widths of the credible intervals from Figure 18.4. Uncertainty is higher in the middle of the 2006–2013 census cycle, which lasted 7 years, than in the middle of the 1996–2001 and 2001–2006 census cycles, which both lasted 5 years. As can be seen in Figure 18.5, uncertainty also grows more quickly after the 2013 census. There is no later census to anchor the estimates.

We do not show estimates for birth and death counts, because they are virtually indistinguishable from the data for registered births and deaths. The data models for registered births and deaths imply that data are highly accurate, and the demographic accounting provides no evidence to the contrary.

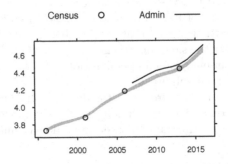

FIGURE 18.4: Population estimates from national demographic account (millions). The gray bands represent 95% credible intervals. The black points represent census-based population estimates, and the black lines represent admin-based population estimates.

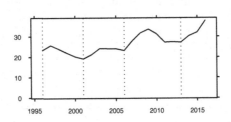

FIGURE 18.5: Widths of 95% credible intervals for population estimates from national demographic account (thousands).

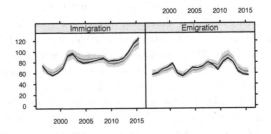

FIGURE 18.6: Estimates of international migration from national demographic account (thousands).

The estimates for immigration and emigration, in contrast to the estimates for birth and death counts, contain substantial uncertainty. As can be seen in Figure 18.6, 95% credible intervals for the series cover a range of about 20,000.

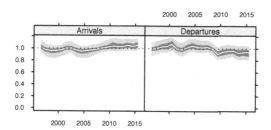

FIGURE 18.7: Coverage ratios for international migration from national demographic account. The coverage ratios show the number of international arrivals or departures divided by the true number of immigrations or emigrations (as estimated within the national demographic account).

Figure 18.7 shows coverage ratios for the arrival and departure data. These coverage ratios are defined differently from the ones we used in the Cambodian chapter (Sections 16.5–16.7). There, we used the parameter γ_{ap}^{cen} from the model for census coverage. Here we use the ratio of observed arrivals to estimated true immigrations, and the ratio of observed departures to estimated true departures. Using the terminology introduced in Section 4.9, the coverage ratio in the Cambodian chapter was a super-population quantity, and the one in Figure 18.7 is a finite-population quantity. The advantage of the finite-population version is that it is not tied to any particular model. The coverage ratios fluctuate over time, but are centered around 1.

18.4 Sensitivity Tests for the National Demographic Account

In our experience, using informative priors is an essential part of the estimating a demographic account. Informative priors in the system model allow us to incorporate, in a formal way, the sort of knowledge about trends and plausible ranges that demographers have always brought to bear on demographic problems. Informative priors and strong assumptions seem to be unavoidable for data models. The framework of Part V allows us to drop the assumption of perfect data, but does not free us from the need to assume *something* about data quality.

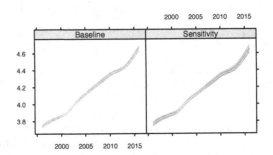

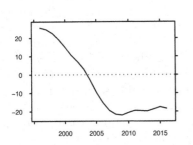

(a) National population estimates (millions)

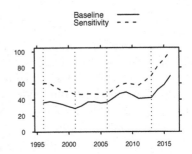

(b) Posterior median of alternative model minus posterior median of baseline model (thousands)

(c) Widths of 95% credible intervals (thousands)

FIGURE 18.8: A sensitivity test for population estimates from the national demographic account. The baseline model assumes that census-based population estimates have a 95% chance of being within 0.5% of the true population counts, and the alternative model assumes that the estimates have a 95% chance of being within 1% of the true population counts.

There is never enough time to check all the assumptions in a model. It is, however, prudent to check assumptions that lack strong justification, or that are likely to have a major effect on the results. To illustrate, in this section we examine the effect of altering the assumption from Section 18.2.4 that census-based estimates are within 0.5% of the true population counts. We replace this with an assumption that the estimates are within 1% of the true population counts.

The results are shown in Figure 18.8. Panel (a) shows population estimates under our original assumption (the "baseline" model) and the alternative assumption (the "alternative" model). The effect of changing from 0.5% to 1% is evidently small. Panel (b) shows differences in the posterior medians. The differences range from 20,000 to -20,000, which is a lot of people, but still a small proportion of a total population of 4–5 million. As can be seen in Panel (c), the alternative assumption implies wider credible intervals, and also less variability in uncertainty over the census cycle.

18.5 Input Data for the Regional Demographic Account

We now add a region dimension to our demographic system and demographic account. For simplicity, however, we continue to ignore age, sex, or any other attributes besides region. We also reduce the size of the account down to 11 time points, from 2006 to 2016.

In our second demographic system, New Zealand is divided into 16 regions. The largest such region, Auckland, had an estimated population in 2013 of 1.5 million, while the smallest region, the West Coast, had a population of 33 thousand. Because people can change their region of residence, we need to extend our account to allow for movements between regions. We also need to disaggregate the existing series—population, births, deaths, immigration, and emigration—by region.

TABLE 18.3
Data sources for internal migration

Data source	Accuracy
Census-based transition	Moderate
Admin-based address change	Moderate

We have two data sources providing information on internal migration, summarized in Table 18.3. The first is census data on the region of New Zealand that the respondent lived in 5 years before the current census. This data allows us to calculate transitions, as defined in Section 4.4. Transitions data on migration measure, for instance, the number of people who were in Auckland in 2008 and in Canterbury in 2013. Transitions measure different aspects of migration from movements, especially when the transitions are calculated over multi-year periods. A person who was in Auckland in 2008 and Canterbury in 2013, for instance, could potentially have made multiple movements between other regions during the period 2008–2013, none of which would be recognized in the transitions data.

The second data source on internal migration is annual address changes, as recorded within the individual-level datasets that were used to calculate admin-based population estimates. If New Zealanders only ever used their residential address in correspondence with government departments, if they notified the government every time they changed residence, and if addresses were always accurately recorded, then annual address changes would measure the annual number of moves perfectly. Unfortunately, none of these conditions are met.

Figure 18.9 shows the census and address change data for five selected origin and destination regions. The regions are ordered from smallest (Gisborne) to largest (Auckland). As with the Icelandic migration data (Section 15.1), the panels on the diagonal are blank, since we are ignoring movements within each region.

The flows vary vastly in size, from thousands (e.g. from Canterbury to Auckland) to dozens (e.g. from Gisborne to Tasman). We therefore use a different vertical scale for each panel. The smaller flows are subject to substantial random variation. Overall, however, the number of 5-year transitions is roughly equal to the annual number of address changes in each panel. This is somewhat surprising, given that the two data sources capture different aspects of migration.

The use of different vertical scales for the panels obscures the strength of the relationship between 5-year transitions and annual address changes. Figure 18.10 shows the relationship more clearly. Apart from a few very small flows (which are subject to substantial random variation) 5-year transitions do a good job of predicting annual address changes, and vice versa. Despite the differences in definition between 5-year transitions and annual address changes, they are both ultimately measures of migration propensities, which presumably explains why there is a strong relationship.

A second striking feature of the data is that address changes are increasing steadily over time. It is possible that the increase reflects real change in migration behavior. However, it is also possible that the increase is an artefact of the way the addresses are collected. For instance, as government IT systems improve, perhaps address changes are picked up more quickly.

For the remaining demographic series, we use the same data sources as we did for the national demographic accounts. All these data sources are disaggregated to the regional level, apart from the admin-based population estimates, which are at the national level. The data are graphed in Figure 18.11.

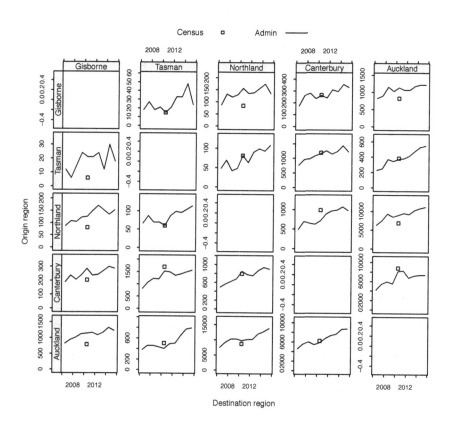

FIGURE 18.9: Data on internal migration for regional demographic account, for five selected regions. Each row of panels shows data for an origin region, and each column shows data for a destination region. The bottom-left panel, for instance, shows data on flows from Auckland to Gisborne. The census data are the number of people living in the destination region who were living in the origin region 5 years ago. The administrative data is based on address changes reported through government agencies. The vertical scales vary from panel to panel.

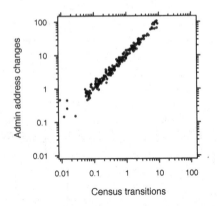

FIGURE 18.10: Administrative address changes versus census 5-year transitions. Each point corresponds to an origin-destination pair. The point shows the average value for annual address changes versus the value for 5-year transitions for that pair. The horizontal and vertical axes both use a log scale.

18.6   Model for the Regional Demographic Account

18.6.1   System Models

The system models for the regional demographic account look much like the system models for the national demographic account, in that they are all simple Poisson models with moderately informative priors on the standard deviation terms. The ability to re-use models and code when moving from one problem to another is one of the advantages of using a formal statistical approach to demographic estimation. The main effects, interactions, and associated priors for population, births, deaths, immigration, and emigration are summarized in Table 18.4.

TABLE 18.4
Main effects, interactions, and priors in system models for regional demographic account

Model	Region	Time	Region-time
Population	Normal	Local trend	Local trend
Births	Normal	Local level	None
Deaths	Normal	Local trend	None
Immigration	Normal	Local level	Local level
Emigration	Normal	Local level	Local level

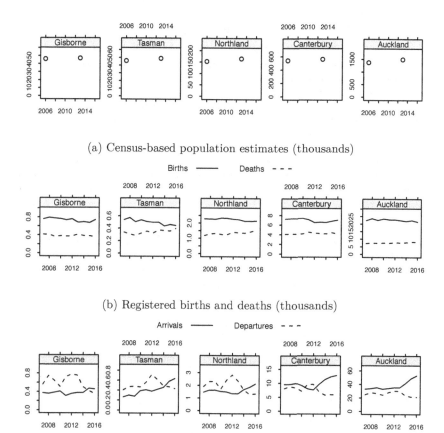

(a) Census-based population estimates (thousands)

(b) Registered births and deaths (thousands)

(c) International arrivals and departures (thousands)

FIGURE 18.11: Input data for regional demographic account (aside from data on internal migration for five selected regions). The regions are ordered by population size, from smallest to largest. The vertical scales vary from panel to panel.

The 'Normal' prior for the regional effect referred to in Table 18.4 is

$$\beta_r^{\text{reg}} \sim \mathrm{N}(0, \tau^2), \tag{18.13}$$

where τ has a half-t prior with scale 1. The prior is sufficiently simple, and the data contains enough information on regional variation, so that we do not bother specifying a more informative prior. The standard deviations in the local level and local trend priors all have half-t priors with scale 0.05.

For internal migration, we use the same type of model as we used for Iceland (Section 15.4). We assume that the number of movements between regions i and j has distribution

$$y_{ijt}^{\text{int}} \sim \begin{cases} \mathrm{Poisson}(\gamma_{ijt}^{\text{int}} w_{it}) & \text{if } i \neq j \\ 0 & \text{if } i = j, \end{cases} \tag{18.14}$$

where w_{it} is the exposure term for region i. This specification allows for structural zeros where the origin region equals the destination region. We fit a prior model with an origin effect, a destination effect, a time effect, and an origin-destination interaction.

Studies of migration in many countries and time periods have found that migration rates (and not just counts) vary with the population size of the origin and destination regions. Because the population sizes of the New Zealand regions vary so much, these effects are potentially important. We therefore include population size as a covariate in our priors for the origin and destination effects.

$$\beta_i^{\text{int, orig}} \sim \mathrm{N}(p_i \psi^{\text{orig}}, \tau_{\text{int, orig}}^2), \tag{18.15}$$

$$\beta_i^{\text{int, dest}} \sim \mathrm{N}(p_i \psi^{\text{dest}}, \tau_{\text{int, dest}}^2), \tag{18.16}$$

where p_i is the logarithm of population of region i in 2013, according to the census-based population estimates. The parameter ψ^{orig} or ψ^{dest} measures the relationship between the covariate and the origin or destination effect. The origin-destination interaction has a normal distribution with mean 0. The standard deviations for the origin effect, destination effect, and origin-destination interaction all have weakly informative priors.

We model time effects using a local level model. We do not include any sort of time-region interaction. Given the uncertain quality of the data on internal migration, it would be over-ambitious to try to model any sort of region-specific time trends.

18.6.2   Data Models

We use the same data models in the regional demographic account that we used in the national demographic account. However, we also add two new models for the two new data sources.

Let x_{ijp}^{tran} denote the census count of 5-year transitions from region i to region j in a 5-year period p. Let y_{ijp}^{int} denote the aggregated true count of movements from region i to region j for the 5-year period preceding the census. The census data on 5-year transitions are modeled using

$$x_{ijp}^{\text{tran}} \sim \begin{cases} \text{Poisson}(\gamma_{ijp}^{\text{tran}} y_{ijp}^{\text{int}}) & \text{if } i \neq j, \\ 0 & \text{if } i = j, \end{cases} \tag{18.17}$$

$$\log \gamma_{ijp}^{\text{tran}} \sim \text{N}(\mu^{\text{tran}}, \sigma_{tran}^2), \quad i \neq j. \tag{18.18}$$

The data model for transitions, like the system model for internal movements, needs to be able to cope with structural zeros.

Equation (18.17) uses aggregated movements to predict transitions. Even with perfect data, we would not expect aggregated movements to precisely predict 5-year transitions, since movements and transitions measure different aspects of the migration process. But, as can be seen in Figure 18.10 there is a strong relationship between address changes and transitions, at least in this particular dataset. Address changes *are* in principle closely related to movements, so the model of Equations(18.17) and (18.18) can be justified on pragmatic grounds.

We use a very weak prior for μ^{tran},

$$\mu^{\text{tran}} \sim \text{N}(0, 10^2), \tag{18.19}$$

to let the data decide on the overall average ratio between transitions and movements. However, we use a half-t prior with scale 0.025 for σ_{tran}, which is a relatively strong prior. This prior implies that we expect the ratio between transitions and movements to be similar across origin-destination pairs. Without making some sort of reasonably strong assumption about similarities between pairs, it is impossible to conclude anything from the data.

Our data model for admin address changes is similar to our model for transitions, except that we use 1-year periods and include a time trend. Let x_{ijt}^{addr} denote the count of admin address changes from region i to region j in year t. The admin address changes are modeled using

$$x_{ijt}^{\text{addr}} \sim \begin{cases} \text{Poisson}(\gamma_{ijt}^{\text{addr}} y_{ijt}^{\text{int}}) & \text{if } i \neq j, \\ 0 & \text{if } i = j, \end{cases} \tag{18.20}$$

$$\log \gamma_{ijt}^{\text{addr}} \sim \text{N}(\beta_0^{\text{addr}} + \beta_t^{\text{time,addr}}, \sigma_{\text{addr}}^2), \quad i \neq j. \tag{18.21}$$

We assume that the overall ratio between address changes and actual movements is changing linearly over time. In other words, the time effect has prior

$$\beta_t^{\text{time,addr}} \sim \text{N}(\alpha_0 + \alpha_1 t, \tau_{\text{addr}}^2). \tag{18.22}$$

We could, in principle, have used a local trend model, which would allow the trend to vary, rather than be fixed across all years. However, the address change data only covers a period of 11 years, and the series in Figure 18.9 seem to share a common linear trend.

18.7   Results for the Regional Demographic Account

Figure 18.12 shows estimates of regional population counts. The credible inter-vals are relatively narrow: the available data apparently permit quite precise estimates, at least at the level of total counts.

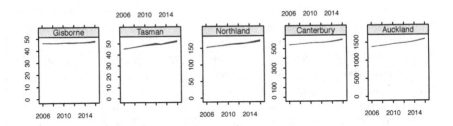

FIGURE 18.12: Estimates of population (thousands) in five selected regions from the regional demographic account. The gray bands represent 95% credible intervals. Each panel has a different vertical scale.

Figure 18.13 shows estimates of the internal migration flows that, along with births, deaths, and internal migration, produce the regional population counts. The estimates are more precise for Auckland than for other regions, reflecting the larger size, and smaller random fluctuations, of the flows into and out of Auckland. Even outside Auckland, however, the posterior distributions change more smoothly from year to year than the address changes.

The estimated number of internal migrations is generally less than the number of address changes, implying that the address change data overstate actual migration. It seems, however, that the relationship between address changes and internal migration has been changing over time. The migration and address changes have both been trending upwards. In many cases, how-ever, the upward trend is steeper for address changes than for the migration estimates. The coverage ratios for the address change data, displayed in Fig-ure 18.14, reinforce this point. With many, though not all, flows, the number of address changes per actual internal migration appears to have been rising.

Why would the model adjust internal migrations downwards from the level suggested by the address change data? In addition to the direct evidence on internal migration provided by the census and address change data, the model has indirect evidence from the data on population, births, deaths, and international migration. When estimates for these quantities are combined via the accounting identities, they provide an alternative estimate of internal migration. It seems that, in balancing the two sources of evidence, the model has adjusted some of the internal migration estimates downwards.

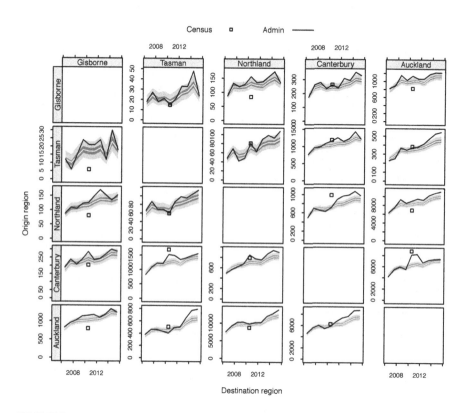

FIGURE 18.13: Estimates of internal migration between five selected regions from the regional demographic account. Each panel has a different vertical scale. The light gray bands represent 95% credible intervals, the dark gray bands represent 50% credible intervals, and the pale lines represent posterior medians. Census data and address change data are repeated from Figure 18.9.

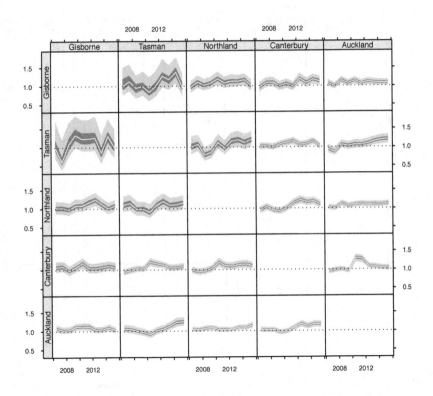

FIGURE 18.14: Coverage ratios for the address change data.

It is not difficult to generate hypotheses for why the relationship between address changes and actual internal migration might have been changing. As correspondence with the government has shifted from paper to the Internet, perhaps the number of distinct addresses recorded within government systems has been increasing.

18.8 References and Further Reading

The population census, vital registration, immigration, and administrative data are from custom tabulations from Stats NZ. All data have been confidentialized. The Stats NZ staff preparing the tabulations accessed the anonymized administrative data in accordance with the security and confidentiality provisions of the Statistics Act 1975. Only people authorized by the Statistics Act 1975 are allowed to see data about a particular person, household, business, or organization.

Bryant and Graham (2013) is an earlier attempt to construct a Bayesian demographic account for the New Zealand population. Toti et al. (2017) apply the same methods to Italian data.

Daponte et al. (1997) use Bayesian methods to estimate the size of the Iraqi Kurdish population. Wheldon et al. (2013) and Wheldon et al. (2015) present and apply a Bayesian approach to simultaneously estimating fertility, mortality, migration, and population, using multiple data sources.

19

Population in China

In our final application, we apply Bayesian demographic accounting to the problem of estimating and forecasting population, births, and deaths in China. We define a relatively simple demographic system, summarized in Table 19.1. We recognize differences by age, but not by any other dimension. We allow people to enter and leave the system via births and deaths, but ignore international migration, which, relative to the vast size of the Chinese population, is a minor contributor to demographic change.

TABLE 19.1

Chinese demographic system modeled in Chapter 19

Membership	Usual resident
Classification	Age, Lexis triangle*
Entries	Births
Exits	Deaths
Movements	Aging

*Births and deaths only

As discussed in Section 4.7, including Lexis triangles in the classification allows us to switch between age and cohort perspectives. As discussed in Section 5.5, the age perspective is important because demographic data and demographic estimates and forecasts are usually disaggregated by age group, rather than by cohort; the cohort perspective is important because demographic accounting identities are based on cohorts.

We consider only three data sources: census counts, existing births estimates, and existing deaths estimates. As we will see, however, Chinese demographic data have some major internal contradictions, with cohorts evolving in ways that are not demographically possible.

We use the framework of Part V to synthesise evidence from the three data sources and construct accounts that are internally consistent. The analyses in this chapter are only illustrative: properly reconstructing the Chinese demographic system would require more than three data sources, and more dimensions than just age. The aim of the chapter is to demonstrate how Bayesian demographic accounts deal with the challenges that arise in practical examples.

19.1   Input Data

Our main data source is (unadjusted) Chinese census data for the years 1990, 2000, and 2010, disaggregated by 5-year age group. The most recent census data are, in principle, accurate. According to a coverage survey carried out after the 2010 census, the census undercounted the true population by only 0.12%. The scholarly concensus is that earlier censuses are less accurate.

The other two datasets are estimates of births and deaths, by 5-year age group, for the periods 1990–1995, 1995–2000, ..., 2010–2015, constructed by United Nations demographers. The UN estimates start with official Chinese figures, but include adjustments based on various sources of evidence, including 2000 and 2010 census data. The original Chinese registration and survey data for births and deaths are not publicly available.

The three datasets are plotted in Figure 19.1. Even allowing for possible errors in the data, the rapid changes in age structure between the three censuses are extraordinary. The number of children aged 0–4, for instance, dropped by tens of millions between 1990 and 2000, reflecting the dramatic fall in Chinese fertility that began in the 1960s.

Although each series in Figure 19.1 looks plausible on its own, when we put them together in an accounting framework, we uncover inconsistencies. To see these inconsistencies, we need to apply some of the principles of demographic accounting introduced in Chapter 5.

Because the census is conducted every 10 years, whereas births and deaths are estimated over every 5-year period, we look at census counts in two years: t and $t + 10$, and estimates of births and deaths in two periods: between t and $t + 5$, and between $t + 5$ and $t + 10$.

Consider a cohort born during the period between $t + 5$ and $t + 10$, as shown in Panel (a) of Figure 19.2. The only sources of population change are births and deaths. In year $t + 10$, members of the cohort born between $t + 5$ and $t + 10$ are aged 0–4. Some members of the cohort have died, so the size of the cohort in year $t + 10$ is less than the number of births.

Similar reasoning applies to the cohort born during the period between t and $t + 5$, shown in Panel (b) of Figure 19.2. At time $t + 10$, members of this cohort are aged 5–9. Since cohort members are lost to death, but not replaced by any other process, the size of the cohort in $t + 10$ is less than the number of births.

Similarly, the cohort that was already born in year t, shown in Panel (c) of Figure 19.2, should be smaller in year $t + 10$ than it was in year t.

Sadly, the census and births data for China do not respect these accounting constraints. Figure 19.3 shows how cohorts constructed from the census and births data develop over time.

The two cohorts at the top left of the graph, which were born in 1950–1955 and 1955–1960, reduce in size over both 10-year periods. The remaining

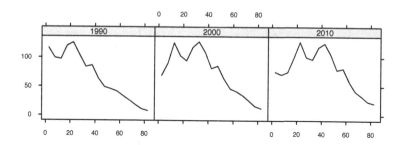

(a) Census counts of population (millions), by age.

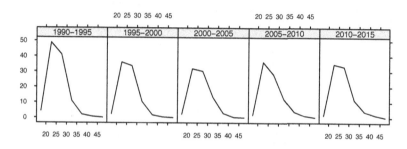

(b) Estimates of births (millions), by age of mother.

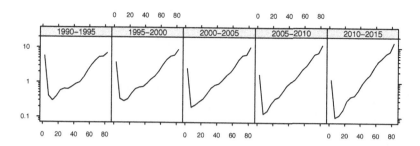

(c) Estimates of deaths (millions), by age. The data are shown on a log scale.

FIGURE 19.1: Input data for the Chinese demographic account.

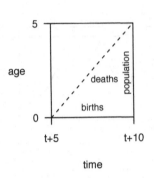

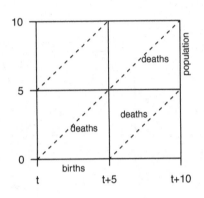

(a) Cohort born between times $t+5$ and $t+10$

(b) Cohort born between times t and $t+5$

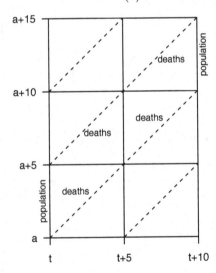

(c) Cohort already born by time t

FIGURE 19.2: Changes in cohort size in a demographic system with population, births, and deaths, described by Lexis diagrams, as defined in Section 3.3.

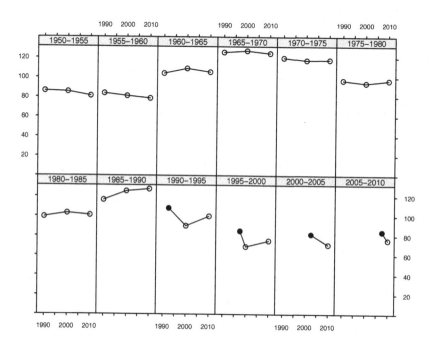

FIGURE 19.3: Changes in cohort size (millions), as implied by the births and census data. Empty circles denote population counts at points in time. Filled circles denote birth counts over 5-year periods, and are plotted at the middle of the corresponding periods. Each panel shows a different cohort. The top-left panel, for instance, shows the cohort born during the period 1950–1955.

cohorts, other than those born in the 2000s, all gain in size over at least one period, though the cohorts born in the 1980s and 1990s show the most dramatic upward movements. Though logically possible, the steep drops in the sizes of the cohorts born in the 1990s and 2000s are also suspicious, as they imply much higher death rates for children than are normally attributed to China over this period.

19.2 Model

19.2.1 Overview

Figure 19.4 summarizes the overall model. The true population counts are classified by age, and the true birth and death counts are classified by age and Lexis triangle. The true counts of population, births, and deaths are linked through accounting identities, as indicated by the big rectangle. As discussed in Section 17.1, exposures for births and deaths are calculated internally using population counts. Each demographic series has one system model, one data source, and one data model.

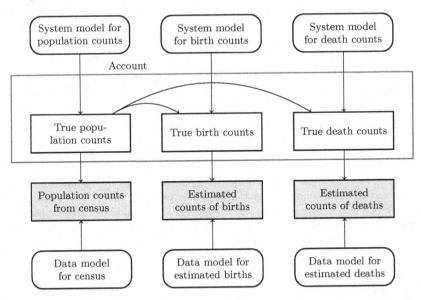

FIGURE 19.4: Overview of demographic account for China. Small straight-edged rectangles represent demographic arrays, and rounded rectangles represent models. The gray rectangles are observed; everything else is inferred. The big rectangle marks out the demographic account.

19.2.2 Account

The account includes true population counts for the years 1990, 1995, ..., 2020, and true counts of births and deaths for the periods 1990–1995, 1995–2000, ..., 2015–2020. Population and death counts are disaggregated by 5-year age groups 0–4, 5–9, $\cdots$, 80+. Birth counts are disaggregated by 5-year age groups 15–19, 20–24, $\cdots$, 45–49. With birth counts, age refers to age of the parents.

Since the input data for population counts do not include years 1995, 2005 and 2015, counts of population in these years are, from the point of view of the model, missing data (Section 9.4). As well as age, the true counts of births and deaths are also disaggregated by Lexis triangle. Since the input data do not include Lexis triangles, counts of births and deaths in each triangle are missing data.

Furthermore, we only have data up to 2015. The last 5 years of the account are, accordingly, forecasts.

19.2.3 System Models

We use Poisson distributions for all the system models. As usual, the model for population does not have an exposure term, while the models for births and deaths do.

Our models for births and deaths include separate rates for upper and lower Lexis triangles, within each combination of age and period. Because our data do not provide any direct information distinguishing upper and lower triangles, we do not include any main effects or interactions involving Lexis triangles in our models for births and deaths. In the system models, the rates for the upper and lower Lexis triangles have the same expected values, but can be different due to random variation.

We also need to have separate exposure measures for upper and lower Lexis triangles. We set exposure for an upper Lexis triangle equal to one half of the population at the start of the period multiplied by the length of the period. Exposure for a lower Lexis triangle is defined as one half of the population at the end of the period multiplied by the length of the period.

The main effects, interactions, and associated priors for population, birth, and death counts are summarized in Table 19.2. We chose the terms to include in the models by carrying out decompositions similar to the ones described in Sections 12.4 and 12.5.

The standard deviations in the models all have weakly informative half-t priors, except for the standard deviations for the level and trend terms involving time in the models for births and deaths. With only a few time periods of data to draw on, estimating these terms is difficult without prior information. We use scales of 0.1 for time effects and 0.05 for age-time interactions.

TABLE 19.2

Main effects, interactions, and priors in system models for
demographic account for China

Model	Age	Time	Age-time
Population	Local trend	Local trend	Local trend
Birth	Local trend	Local level	Local level
Death	Local trend	Local trend	Local level

19.2.4   Data Models

We use x_{at}^{cen} to denote census count for age group a at time t, and use y_{at}^{popn}
to denote the true population count. We assume that errors in the census are
normally distributed with a mean of zero, so that

$$x_{at}^{\text{cen}} \sim \text{N}(y_{at}^{\text{popn}}, s_{at}^2). \tag{19.1}$$

We set s_{at} to $0.005 \times x_{at}^{\text{cen}}$, which implies that the census count x_{at}^{cen} has an
approximately 95% chance of being within 1% of the true population count
y_{at}^{popn} (see Section 18.2.4 for a related discussion). This is more precise than
would be suggested by our analysis of changes in cohort size, but less precise
than would be suggested by the official undercount estimate of 0.12%.

Counts of estimated births receive a similar model to the one we use for
the admin-based population counts in Section 18.2.4. For age group a and
time period t, let x_{at}^{bth} denote the UN estimates of births, and let y_{at}^{bth} denote
the true count. The data model is

$$x_{at}^{\text{bth}} \sim \text{Poisson}(\gamma_{at}^{\text{bth}} y_{at}^{\text{bth}}), \tag{19.2}$$

$$\log \gamma_{at}^{\text{bth}} \sim \text{N}(\mu^{\text{bth}}, \sigma_{\text{bth}}^2). \tag{19.3}$$

The coverage ratio γ_{at}^{bth} is allowed to vary over age group and over time.

The mean parameter μ^{bth} controls the average coverage ratio across age
groups and time periods. We set a moderately informative prior

$$\mu^{\text{bth}} \sim \text{N}(0, 0.1^2). \tag{19.4}$$

This implies that we expect the coverage ratios to be centered at a level somewhere between 0.8 and 1.2. The parameter σ_{bth} governs the variability of the
coverage ratios around their overall average. We use a moderately informative
prior for σ_{bth}, a half-t distribution with a scale of 0.1. This allows the coverage
ratio for a specific age group and a specific time period to be different from
the overall average by 0.2.

The data model for death counts is identical to the one for birth counts.

19.2.5   Estimation and Forecasting

As discussed in Section 9.5, forecasting can be approached as a missing data
problem. We treat population counts at 2020, and birth and death count

during 2015–2020 as missing data, and estimate them together with other missing data and the unknown parameters.

As with the estimation of the New Zealand account in Section 18.2.5, we iterate through each component of the model, updating that component conditional on current values for the other components. However, since in this case our account involves age, updating a population, birth or death count involves tracking a cohort across periods, and making sure that no counts calculated using the accounting identities are non-negative.

Suppose, for instance, that we want to update the death count in the upper Lexis triangle for age 15–19 during the period 1990–1995, conditional on current values for the other components. Figure 19.5 shows the population counts in later years that will be affected by the changes, as well as the death counts experienced by the cohort. If the updated death count is higher than the current one, then subsequent population counts will be lower. We need to make sure that the updated death count is not so high that subsequent population counts become negative.

Although the accounting is done cohort-by-cohort, the system models take an age perspective, with age effects but not cohort effects. As noted above, including Lexis triangles in the account allows us to work simultaneously with age groups and cohorts.

19.3 Results

We look first at the results for population, birth, and death counts. Figure 19.6 shows estimated and forecasted counts, as well as the original census, birth, and death data.

The population estimates in Panel (a) are close to the census counts, though some differences are discernable at the younger ages in 2000 and 2010. The population forecasts for 2020 have substantial uncertainty at the youngest and oldest ages.

The estimates for birth counts, shown in Panel (b), include a Lexis triangle dimension. The top row shows estimates for upper Lexis triangles, and the bottom row shows estimates for lower triangles. The original UN counts do not include Lexis triangles. To allow for comparison between the UN counts and our estimates, we distribute the UN counts evenly between upper and lower triangles. There is considerable uncertainty about birth counts, particularly in the forecast period 2015–2020. Our estimates are also somewhat lower than the UN ones.

Our estimates for death counts, shown in Panel (c), also include Lexis triangles. Once again, we distribute the UN counts, which do not include Lexis triangles, evenly between upper and lower triangles. Overall, our estimates

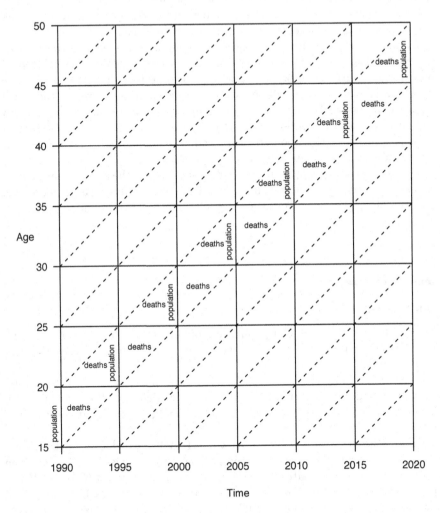

FIGURE 19.5: Updating a cohort within the account.

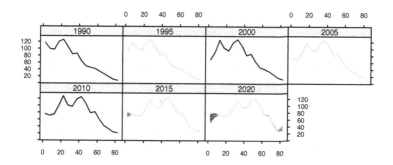

(a) Population counts

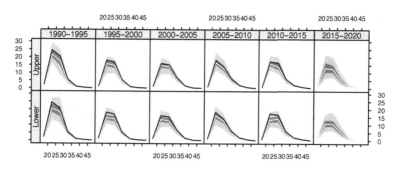

(b) Birth counts

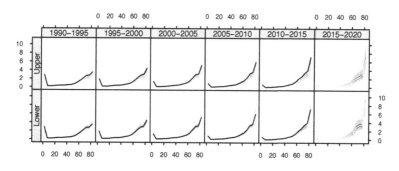

(c) Death counts

FIGURE 19.6: Estimated and forecasted population, birth, and death counts. The light gray bands represent 95% credible intervals, the dark gray bands represent 50% credible intervals, and the pale lines represent posterior medians. The black lines are census counts and UN estimates.

agree with the UN numbers, though there are some differences at the oldest and youngest age groups.

Figure 19.7 shows the posterior means for cohort sizes, together with the original unadjusted sizes. We use posterior means rather than medians because the means satisfy the accounting identities while the medians do not. (The mean is a 'linear' function of its inputs, while the median is not.) As we would hope, the posterior means from the model respect the accounting constraints. No cohort increases in size over time. Furthermore, the sharp drops in cohort size immediately after birth have gone, implying much more plausible levels of childhood mortality.

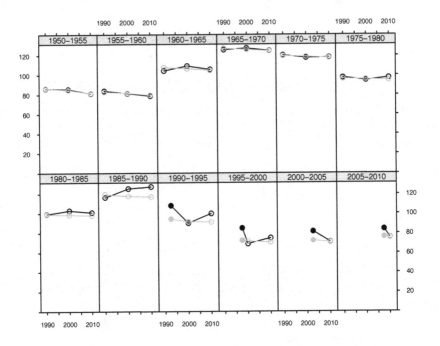

FIGURE 19.7: Changes in cohort size (in millions), from the input data and demographic account. Black denotes input data and gray denotes posterior means. Empty circles denote population counts at points in time. Filled circles denote birth counts over 5-year periods, and are plotted at the middle of the corresponding periods. Each panel shows a different cohort.

As well as counts from the demographic account, we obtain super-population quantities from the system models. Figure 19.8a, for instance, shows estimated and forecasted super-population death rates. (We show a weighted average of the death rates for upper and lower Lexis triangles, with the exposures for the triangles as weights.) Forecasted death rates for 2015–2020 are much less certain than estimated death rates for earlier periods.

The estimated and forecasted death rates can be converted into estimated and forecasted life expectancies, as shown in Figure 19.8b. The 95% credible interval for the forecasting period is wide, reflecting the large uncertainty in the model parameters. The uncertainty in the parameters partly reflects uncertainties in the input data: unlike in Part III, the deaths and exposures used to calculate historical rates are not treated as known. But the uncertainty is also a consequence of using a historical time series with only five periods.

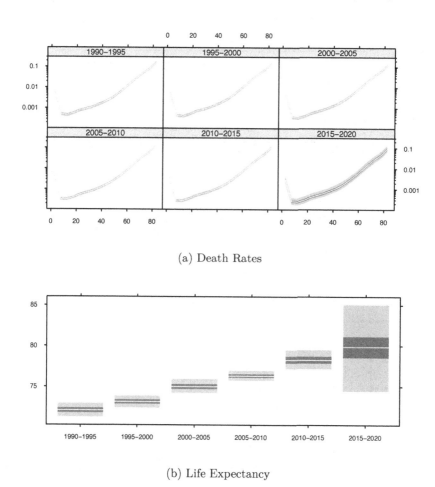

(a) Death Rates

(b) Life Expectancy

FIGURE 19.8: Estimated and forecasted death rates and life expectancy. The light gray bands represent 95% credible intervals, the dark gray bands represent 50% credible intervals, and the pale lines represent posterior medians.

The last set of estimates represents coverage ratios. Figure 19.9 shows ratios of counts from the input data to estimated true counts from the associated demographic series.

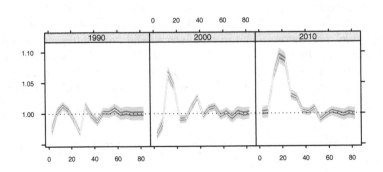

(a) Census counts

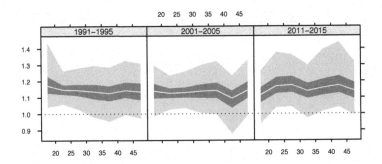

(b) UN births estimates

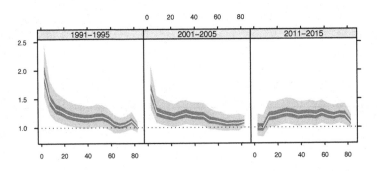

(c) UN deaths estimates

FIGURE 19.9: Estimated coverage ratios from the Chinese demographic account. The light gray bands represent 95% credible intervals, the dark gray bands represent 50% credible intervals, and the pale lines represent posterior medians.. To save space, coverage ratios for births and deaths in 1995–2000 and 2005–2010 have been omitted.

The coverage ratios for the census, in Panel (a), are around 1 for the older age groups, but vary for the younger age groups. The age profile for coverage is similar for the 2000 and 2010 census, but not the 1990 census.

There is substantial uncertainty about coverage ratios for births, shown in Panel (b), though the overall message is that the UN birth estimates are higher than ours. The difference between the UN estimates and our estimates is even greater for deaths, shown in Panel (c), especially for the youngest age groups. This is in line with the analysis in Section 19.1 which demonstrates that the input data imply too many deaths for children.

19.4   References and Further Reading

The Chinese census data come from the table *Population by age, sex and urban/rural residence* in the United Nations Statistics Division online database *Population Censuses' Datasets (1995 - Present)*, and were downloaded on January 21, 2018. The figure of a 0.12% undercount comes from the footnotes from the census dataset. The estimates of births and deaths come from the tables *Number of births by age of mother (thousands)* and *Number of deaths by age and sex (thousands)* in the United Nations Population Division online database *World Population Prospects*. The data were downloaded on November 30, 2017 and December 2, 2017.

Cai (2013) is full-scale analysis of the 2010 Chinese census data, bringing in far more data sources than we consider in this chapter. Zhang and Zhao (2006) is a similarly comprehensive analysis of Chinese fertility data.

20

Conclusion

In this book, we have examined three frameworks for demographic estimation and forecasting. The first framework is the simplest and most mature. The third is the most complicated and least mature, but has, we hope, a bright future.

The first framework deals with the case where we have a single demographic series, and our data are reliable. The main methodological challenges are disaggregation and forecasting.

We deal with the challenges of disaggregation by setting up models that account for random variation, and that share information across the dataset as a whole. It turns out that, to capture all the trends and counter-trends in big demographic datasets, we need to fit complicated models, with many interactions. There are, however, ways of keeping this complexity manageable. We can build a model piece by piece, using decompositions and graphs to guide the construction of each piece.

Throughout the book, we have emphasized that forecasting is a type of estimation. That does not imply, however, that forecasting is easy. Forecasts are sensitive to modeling assumptions, and model-checking is essential. Assembling long data series for disaggregated forecasts can be difficult, though judicious use of informative priors can compensate for limited data in a principled and transparent way.

Our R packages make it relatively easy to set up, run, and scrutinize the models from this book. We routinely run models with thousands or tens of thousands of parameters on desktop computers. The next step is to move the calculations on to the cloud, and to try models with hundreds of thousands, or millions, of parameters.

We have tried models from the first framework on many different datasets, and have used the results to fine tune the methods and software. For many standard problems involving the estimation of disaggregated demographic rates, it is now possible to produce reasonable models quickly and easily.

The second framework goes beyond the first framework by allowing for multiple unreliable datasets. To accommodate unreliable datasets, we need data models. Many of the lessons learned about systems models in the first framework carry through to data models in the second framework. For instance, modeling variation over age when estimating census coverage ratios is a lot like modeling variation over age when estimating mortality rates.

At the same time, however, each dataset has its own characteristic strengths and weaknesses, which need to be represented in the associated data model. Over the long term, we would like to develop a large suite of data models, which analysts can pick and choose from, depending on the datasets to hand. We plan, for instance, to develop data models for cases where some some dimensions are measured more accurately than others, such as when age is measured more accurately than location.

We also plan to construct data models based on the set of techniques that demographers refer to as "indirect methods". These techniques provide ways of detecting and adjusting for characteristic errors in demographic data, particularly data from developing countries. Many of these techniques have proven their worth over decades of use. Most do not, however, provide explicit measures of uncertainty.

The third framework takes the further step of allowing for multiple demographic series, linked by accounting identities. Because the third framework builds on the first two frameworks, lessons learned from the first two again carry through to the third. We have not, however, had as much experience working with the third framework as we have with the other two. Moreover, estimating demographic accounts is not always easy. It can be difficult, for instance, to know how much flexibility to build into the data models, while still retaining enough structure for the model to converge.

Overall, however, our experiences with accounting models have been encouraging. The models do what they are supposed to. Tasks that are conventionally treated separately, such as data evaluation, population estimation, estimation of super-population rates, and demographic forecasting, are all combined into a single process. Uncertainty in one part of the process flows through to the other parts of the process. Estimates of the demographic series, and their underlying rates, are all consistent with each other.

Modeling national populations, disaggregated by dimensions such as age, sex, and region, as we do in Chapters 18 and 19, is the traditional application for demographic accounts, and an important one. We intend to continue refining national population models, and increasing the level of disaggregation.

At the same time, however, we are looking forward to experimenting with non-traditional applications. Potential applications include, for instance, modeling disease prevalence, forecasting future labor supply, and studying promotion within organizations. We hope that readers of this book will try the models out on applications of their own.

20.1 References and Further Reading

Preston et al. (2001, ch. 11) provide an overview of indirect methods, and Moultrie et al. (2013) go into the details.

Bibliography

Abel, G. J., Barakat, B., Samir, K., and Lutz, W. (2016). Meeting the sustainable development goals leads to lower world population growth. *Proceedings of the National Academy of Sciences*, 113(50):14294–14299.

Alexander, M. and Alkema, L. (2018). Global estimation of neonatal mortality using a Bayesian hierarchical splines regression model. *Demographic Research*, 38(15):335–372.

Alho, J. and Spencer, B. (2005). *Statistical Demography and Forecasting*. Springer-Verlag.

Alkema, L., Raftery, A. E., Gerland, P., Clark, S. J., and Pelletier, F. (2012). Estimating trends in the total fertility rate with uncertainty using imperfect data: Examples from West Africa. *Demographic Research*, 26(15).

Asian Development Bank (2014). *Cambodia Country Poverty Analysis 2014*. Asian Development Bank, Manila.

Bijak, J. (2010). *Forecasting International Migration in Europe: A Bayesian View*, volume 24. Springer-Verlag, Dordrecht.

Bijak, J., Alberts, I., Alho, J., Bryant, J., Buettner, T., Falkingham, J., Forster, J. J., Gerland, P., King, T., Onorante, L., Keilman, N., O'Hagan, A., Owens, D., Raftery, A., Ševčíková, H., and Smith, P. W. (2015). Probabilistic population forecasts for informed decision-making. *Journal of Official Statistics*, 31(4):537–544.

Bijak, J. and Bryant, J. (2016). Bayesian demography 250 years after Bayes. *Population Studies*, 70(1):1–19.

Booth, H. and Tickle, L. (2008). Mortality modelling and forecasting: A review of methods. *Annals of Actuarial Science*, 3(1-2):3–43.

Braaksma, B. and Zeelenberg, K. (2015). "Re-make/Re-model": Should big data change the modelling paradigm in official statistics? *Statistical Journal of the IAOS*, 31(2):193–202.

Bryant, J. and Graham, P. (2015). A Bayesian approach to population estimation with administrative data. *Journal of Official Statistics*, 31(3):457–487.

Bryant, J. and Zhang, J. L. (2016). Bayesian forecasting of demographic rates: Emigration rates by age, sex, and region in New Zealand, 2014-2038. *Statistica Sinica*, 26:1337–1364.

Bryant, J. R. and Graham, P. J. (2013). Bayesian demographic accounts: Subnational population estimation using multiple data sources. *Bayesian Analysis*, 8(3):591–622.

Buonaccorsi, J. P. (2010). *Measurement Error: Models, Methods, and Applications*. CRC Press, Boca Raton.

Cai, Y. (2013). China's new demographic reality: Learning from the 2010 census. *Population and Development Review*, 39(3):371–396.

Coleman, D. (2013). The twilight of the census. *Population and Development Review*, 38(s1):334–351.

Courgeau, D. (1985). Interaction between spatial mobility, family and career lifecycle: A French survey. *European Sociological Review*, 1(2):139–162.

Courgeau, D. (2012). *Probability and Social Science: Methodological Relationships between the Two Approaches*, volume 10. Springer-Verlag, Dordrecht.

Cutler, D. M. and Sheiner, L. (1998). Demographics and medical care spending: Standard and non-standard effects. Technical report, National Bureau of Economic Research, Cambridge, MA.

Daponte, B., Kadane, J., and Wolfson, L. (1997). Bayesian demography: Projecting the Iraqi Kurdish population, 1977-1990. *Journal of the American Statistical Association*, 92(440):1256–1267.

Dieleman, J. L., Squires, E., Bui, A. L., Campbell, M., Chapin, A., Hamavid, H., Horst, C., Li, Z., Matyasz, T., Reynolds, A., et al. (2017). Factors associated with increases in US health care spending, 1996-2013. *Journal of the American Medical Association*, 318(17):1668–1678.

Fischhoff, B. (2012). Communicating uncertainty: Fulfilling the duty to inform. *Issues in Science and Technology*, 28(4):63–70.

Gelman, A. (2006). Prior distributions for variance parameters in hierarchical models. *Bayesian Analysis*, 1(3):515–533.

Gelman, A., Carlin, J., Stern, H., , Dunson, D. B., Vehtari, A., and Rubin, D. (2014). *Bayesian Data Analysis. Third Edition*. Chapman and Hall, New York.

Gelman, A. and Hill, J. (2007). *Data Analysis Using Regression and Multilevel/Hierarchical Models*. Cambridge University Press, Cambridge.

Gelman, A., Jakulin, A., Pittau, M. G., and Su, Y.-S. (2008). A weakly informative default prior distribution for logistic and other regression models. *The Annals of Applied Statistics*, 4(2):1360–1383.

Gerland, P., Raftery, A. E., Ševčíková, H., Li, N., Gu, D., Spoorenberg, T., Alkema, L., Fosdick, B. K., Chunn, J., Lalic, N., et al. (2014). World population stabilization unlikely this century. *Science*, 346(6206):234–237.

Gneiting, T. and Raftery, A. E. (2007). Strictly proper scoring rules, prediction, and estimation. *Journal of the American Statistical Association*, 102(477):359–378.

Gustafson, P. (2003). *Measurement Error and Misclassification in Statistics and Epidemiology: Impacts and Bayesian Adjustments*. CRC Press, Boca Raton.

Hájek, A. (2012). Interpretations of probability. In Zalta, E. N., editor, *The Stanford Encyclopedia of Philosophy*. Stanford University, Winter 2012 edition.

Hoff, P. D. (2009). *A First Course in Bayesian Statistical Methods*. Springer-Verlag, Dordrecht.

Hundepool, A., Domingo-Ferrer, J., Franconi, L., Giessing, S., Nordholt, E. S., Spicer, K., and De Wolf, P.-P. (2012). *Statistical Disclosure Control*. John Wiley & Sons.

Hyndman, R., Koehler, A. B., Ord, J. K., and Snyder, R. D. (2008). *Forecasting with Exponential Smoothing: The State Space Approach*. Springer-Verlag, Berlin.

Hyndman, R. J. and Khandakar, Y. (2008). Automatic time series forecasting: The forecast package for R. *Journal of Statistical Software*, 26(3):1–22.

Kronenberg, T. (2009). The impact of demographic change on energy use and greenhouse gas emissions in Germany. *Ecological Economics*, 68(10):2637–2645.

Lee, R. and Miller, T. (2001). Evaluating the performance of the Lee-Carter method for forecasting mortality. *Demography*, 38(4):537–549.

Lee, R. and Miller, T. (2002). An approach to forecasting health expenditures, with application to the US Medicare system. *Health Services Research*, 37(5):1365–1386.

Lee, R. D. and Carter, L. R. (1992). Modeling and forecasting US mortality. *Journal of the American Statistical Association*, 87(419):659–671.

Little, R. and Rubin, D. (2002). *Statistical Analysis with Missing Data. Second Edition*. John Wiley & Sons, New York.

Little, R. J. (2012). Calibrated Bayes, an alternative inferential paradigm for official statistics. *Journal of Official Statistics*, 28(3):309.

Lomax, N., Norman, P., Rees, P., and Stillwell, J. (2013). Subnational migration in the United Kingdom: Producing a consistent time series using a combination of available data and estimates. *Journal of Population Research*, 30(3):265–288.

Lynch, S. M. (2007). *Introduction to Applied Bayesian Statistics and Estimation for Social Scientists*. Springer-Verlag, New York.

McElreath, R. (2016). *Statistical Rethinking: A Bayesian Course with Examples in R and Stan*. CRC Press, Boca Raton.

Moultrie, T., Dorrington, R., Hill, A., Hill, K., Timæus, I., and Zaba, B. (2013). *Tools for Demographic Estimation*. International Union for the Scientific Study of Population, Paris.

National Institute of Statistics, D. and ICF Macro (2011a). *Cambodia Demographic and Health Survey 2010*. National Institute of Statistics, Directorate General for Health, and ICF Macro.

National Institute of Statistics, D. and ICF Macro (2011b). Cambodia Demographic and Health Survey 2010 [Dataset] khir61fl.dta.

O'Hagan, A., Buck, C. E., Daneshkhah, A., Eiser, J. R., Garthwaite, P. H., Jenkenson, D. J., Oakley, J. E., and Rakow, T. (2006). *Eliciting Experts' Probabilities*. John Wiley & Sons, Boca Raton.

O'Hagan, T. (2004). Dicing with the unknown. *Significance*, 133(September):132–133.

Pfeffermann, D. (2013). New important developments in small area estimation. *Statistical Science*, 28(1):40–68.

Prado, R. and West, M. (2010). *Time Series: Modeling, Computation, and Inference*. CRC Press, Boca Raton.

Preston, S., Heuveline, P., and Guillot, M. (2001). *Demography: Modelling and Measuring Population Processes*. Blackwell, Oxford.

Preston, S. H. and Coale, A. J. (1982). Age structure, growth, attrition, and accession: A new synthesis. *Population Index*, 48(2):217–259.

R Core Team (2016). *R: A Language and Environment for Statistical Computing*. R Foundation for Statistical Computing, Vienna, Austria.

Raftery, A. E., Alkema, L., and Gerland, P. (2014). Bayesian population projections for the United Nations. *Statistical Science*, 29(1):58.

Rao, J. N. and Molina, I. (2015). *Small Area Estimation*. John Wiley & Sons.

Raymer, J., Wiśniowski, A., Forster, J. J., Smith, P. W., and Bijak, J. (2013). Integrated modeling of European migration. *Journal of the American Statistical Association*, 108(503):801–819.

Rees, P. (1985). Population structure and models. In Woods, R. and Rees, P., editors, *Choices in the Construction of Regional Population Projections*. Allen and Unwin, London.

Rees, P. and Willekens, F. (1986). Data and accounts. In Rogers, A. and Willekens, F., editors, *Migration and Settlement: A Multiregional Comparative Study*, pages 19–58. Reidel Press.

Rees, P., Wohland, P., Norman, P., and Boden, P. (2012). Ethnic population projections for the UK, 2001–2051. *Journal of Population Research*, 29(1):45–89.

Rees, P. H. and Wilson, A. G. (1977). *Spatial Population Analysis*. Hodder Arnold.

Rogers, A. (1990). Requiem for the net migrant. *Geographical Analysis*, 22(4):283–300.

Rogers, A. (1995). *Multiregional Demography: Principles, Methods, and Extensions*. John Wiley & Sons.

Rubin, D. B. (1984). Bayesianly justifiable and relevant frequency calculations for the applied statistician. *The Annals of Statistics*, 12(4):1151–1172.

Sarkar, D. (2008). *Lattice: Multivariate Data Visualization with R*. Springer-Verlag, New York.

Schoen, R. (1988). *Modeling Multigroup Populations*. Plenum, New York.

Shang, H. L. (2015). Forecast accuracy comparison of age-specific mortality and life expectancy: Statistical tests of the results. *Population Studies*, 69(3):317–335.

Shmueli, G. (2010). To explain or to predict? *Statistical Science*, 25(3):289–310.

Siegel, J. S. (2003). *Applied Demography: Applications to Business, Government, Law and Public Policy*. Academic Press, San Diego.

Smith, S. K. and Swanson, D. A. (1998). In defense of the net migrant. *Journal of Economic and Social Measurement*, 24(3, 4):249–264.

Smith, S. K., Tayman, J., and Swanson, D. A. (2013). *A Practitioner's Guide to State and Local Population Projections*. Springer-Verlag, Dordrecht.

Stigler, S. M. (1980). Stigler's law of eponymy. *Transactions of the New York Academy of Sciences*, 39(1 Series II):147–157.

Stone, R. (1984). The accounts of society. In *Nobel Prize in Economics Documents*. Nobel Prize Committee.

Swanson, D. A. and Tayman, J. (2012). *Subnational Population Estimates*, volume 31. Springer-Verlag, Dordrecht.

Tantau, T. (2008). The TikZ and PGF packages.

Tetlock, P. E. and Gardner, D. (2015). *Superforecasting: The Art and Science of Prediction*. Penguin Random House, London.

Toti, S., Lipsi, R. M., and Giavante, S. (2017). Regional population estimation with Italian administrative data: Preliminary results of the Bryant and Graham approach.

Tukey, J. W. (1962). The future of data analysis. *The Annals of Mathematical Statistics*, 33(1):1–67.

Tuljapurkar, S. (2013). *Population Dynamics in Variable Environments*. Springer-Verlag, Berlin.

UN Inter-agency Group for Child Mortality Estimation (2013). Levels & Trends in Child Mortality. Report 2013. United Nations Children's Fund, New York, available from www.childmortality.org.

UN Inter-agency Group for Child Mortality Estimation (2014). Levels & Trends in Child Mortality. Report 2014. United Nations Children's Fund, New York, available from www.childmortality.org.

UN Inter-agency Group for Child Mortality Estimation (2015). Levels & Trends in Child Mortality. Report 2015. United Nations Children's Fund, New York, available from www.childmortality.org.

UN Inter-agency Group for Child Mortality Estimation (2017). Levels & Trends in Child Mortality. Report 2017. United Nations Children's Fund, New York, available from www.childmortality.org.

UN Population Division (2015). World Population Prospects: The 2015 Revision, Methodology of the United Nations Population Estimates and Projections. Working Paper ESA/P/WP.242, United Nations, Department of Economic and Social Affairs, Population Division.

United Nations General Assembly (2015). Transforming our world: The 2030 agenda for sustainable development.

Vandeschrick, C. (2001). The Lexis diagram, a misnomer. *Demographic Research*, 4:97–124.

Wachter, K. W. (2014). *Essential Demographic Methods*. Harvard University Press, Boston.

Wegner, P. and Reilly, E. D. (2003). Data structures. In *Encyclopedia of Computer Science*, pages 507–512. John Wiley & Sons, Chichester.

Wheldon, M. C., Raftery, A. E., Clark, S. J., and Gerland, P. (2013). Reconstructing past populations with uncertainty from fragmentary data. *Journal of the American Statistical Association*, 108(501):96–110.

Wheldon, M. C., Raftery, A. E., Clark, S. J., and Gerland, P. (2015). Bayesian reconstruction of two-sex populations by age: Estimating sex ratios at birth and sex ratios of mortality. *Journal of the Royal Statistical Society: Series A (Statistics in Society)*, 178(4):977–1007.

Wild, S., Roglic, G., Green, A., Sicree, R., and King, H. (2004). Global prevalence of diabetes: Estimates for the year 2000 and projections for 2030. *Diabetes Care*, 27(5):1047–1053.

Willekens, F. (2006). Description of the multistate projection model (multistate model for biographic analysis and projection). Technical report, Netherlands Interdisciplinary Demographic Institute.

Wilson, T. and Bell, M. (2004). Comparative empirical evaluations of internal migration models in subnational population projections. *Journal of Population Research*, 21(2):127–160.

Zhang, G. (2014). Exploring Methods to Estimate the Intercensal Population of Aboriginal and Torres Strait Islander Australians. Technical report, Australian Bureau of Statistics.

Zhang, G. and Zhao, Z. (2006). Reexamining China's fertility puzzle: Data collection and quality over the last two decades. *Population and Development Review*, 32(2):293–321.

Index

Printed in the United States
by Baker & Taylor Publisher Services